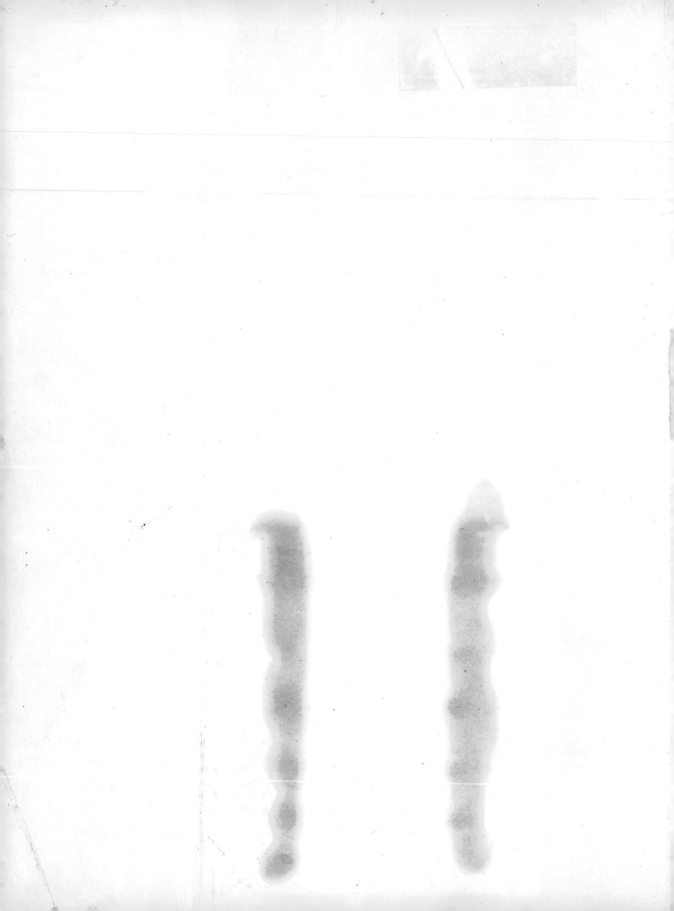

Speed in the Air

Overleaf: The ultimate in operational biplane development, the Gloster Gladiator was the last biplane to enter service on operational duties with the R.A.F. Many were in service during the early years of World War II to struggle valiantly against overwhelming odds – in vain in Norway, but with success in Malta, where three Gladiators, 'Faith', 'Hope' and 'Charity', held the fort until Spitfire reinforcements could come. (Photo: Flight International.)

Also by David Wragg
World's Air Fleets
World's Air Forces
Flight before Flying
A Dictionary of Aviation

SPEED IN THE AIR
DAVID W. WRAGG

A World of Books That Fill a Need

Frederick Fell Publishers, Inc.   New York

Manufactured and first published in England by:
Osprey Publishing Ltd.
707 Oxford Road
Reading, Berkshire

Designed by Behram Kapadia
Filmset and printed by BAS Printers Limited
Wallop, Hampshire

For information address:
Frederick Fell Publishers, Inc.
386 Park Avenue South
New York, NY 10016

In Canada:
George J. McLeod, Limited
Toronto 2B, Ontario

Library of Congress Cataloging in Publication Data
Wragg, David W.
  Speed in the Air.
  1. Airplanes – Speed records. I. Title.
TL 537. W7 1975   629. 13'09   74-7456
ISBN 0-8119-0246-3

Contents

INTRODUCTION

Speed? – definitions and background to powered flight – rules
of the game.

In addition to being a source of fascination for many, the significance of speed in the air lies in the picture it paints of the development of the aeroplane at any stage in its history of little more than seventy years. Speed is, after all, the forte of the aeroplane, and too little has brought more than one 'flying machine' down to earth with considerable force! But the idea of speed in the air is, and always has been, divisible into commercial and military speed as well as the absolute speed record – an event which, before World War I, was connected with the fashionable aviation meetings, between the wars with the sporting air races, and since World War II with the serious technology of the space age.

Obviously, speed is a fairer indication of performance than such criteria as range, endurance, payload and altitude. Range and payload can be offset against each other, while engines throttled back can give a remarkable endurance, but this is failing to use the aircraft for progress through the air and is in fact only a practicable ploy for maritime-reconnaissance or airborne-early-warning duties. Altitude is misleading because there are practical limitations on the heights to be reached for satisfactory performance, and the advent of the space rocket has tended to make this an outmoded concept.

The really significant speed must be the absolute speed, rather than cruising speed, although the latter is the important factor for an airliner. It must also be speed over the ground, rather than speed through the air. Airspeed can be faster or slower than the speed over the ground, but only the latter is meaningful. The introduction of Mach numbers cannot affect this, as the speed of Mach 1.0 is so dependent on local barometric pressure, including the effects of altitude.

For the pioneers, the argument was not really over whether speed or altitude, range or payload was important, but over the definition of flight. The need was to distinguish between flight and powered leaps or hops, in which the force which counted was the take-off run of the aircraft. Flight requires that the engine should be able to sustain the aircraft's progress through the air after the effects of the take-off run have disappeared, and that the endurance of the flight should be long enough for control to be exercised by the pilot.

Wilbur Wright once estimated that it would be possible for a powered leap to extend as far as 250 feet, but some later experts have suggested that as much as a quarter of a mile might be possible.

In its origin, the aeroplane was very much the brainchild of an Englishman, Sir George Cayley (1773–1856), who evolved the correct shape for the aeroplane, and built and successfully tested both model and manned gliders to prove his theories. However, in spite of his attention to the problems of propulsion, and the need for a suitable high output, low weight, power unit, Cayley did not attempt to build a powered flying machine, even in model form.

The construction of the first powered heavier-than-air flying machine was left to another Englishman, W. S. Henson, and his friend, John Stringfellow. They built a steam-powered model which trundled along an overhead launching wire and made a powered glide to the ground. This was in 1842, and although the machine was based on many of Cayley's principles, it did not inspire Cayley to attempt to repeat or improve upon Henson's machine. Henson, for his part bitterly disillusioned, sought such solace as he could in marriage and emigration.

The first full-sized machine to leave the ground was built by a French navy captain, Félix du Temple, in 1874. A sailor drove the steam-powered Du Temple craft down a ramp into what can best be described as a powered leap; it was certainly too short to justify the description of flight although there were those eager to present it as such. A Russian, A. F. Mozhaiski, built a steam-powered machine ten years later, which gave a similar performance at St Petersburg. This too has been spuriously claimed as the first flight; it has even received official recognition in Soviet Russia as such. The Mozhaiski machine is possibly notable for its steam engine being Britain's first aero-engine export, although conspicuously ranking amongst the least successful!

Perhaps more significant than the Du Temple and Mozhaiski attempts was that of another Frenchman, Clement Ader, in 1890. Ader's 'Éole' was the first aeroplane to take-off entirely under its own power, without any down-ramp or down-hill run. Even so the engine was unable to sustain flight. Much of the problem with the 'Éole' can be attributed to Ader's persistence in using the already discredited steam engine for propulsion, as well as the equally dated 'bat wing' concept. The aircraft also showed Ader's railway origins, with the pilot's only view forward being around the boiler, railway-fashion. Ader later built even less successful designs, which after the Wright Brothers' flights, he was to claim as successful. In doing so Ader tarnished his own reputation.

There were other powered flight pioneers, of course, including Horatio Phillips and Thomas Moy, who built successful tethered test rigs, and the American-born Sir Hiram Maxim, who built a large steam-powered test rig in England. The Maxim test-rig actually lifted three men off the ground but, lacking any means of control and kept to a strict course by guide rails, it could do little more with its achievement than foul the guide rails.

The Wright brothers adopted an approach to the subject which was completely different from that of their predecessors. They studied the work of others before attempting to build their own gliders, which acted almost as prototypes for their first aeroplane. In this meticulous manner they steered clear of discredited concepts and were able to evaluate their theories at less cost and risk than many who held themselves in greater self-esteem.

At thirty-five minutes past ten on the morning of 17 December 1903, history saw the first powered flight as Orville Wright took off in the Wright 'Flyer' into a 25 mph wind at 30 mph, to stay airborne for twelve seconds, at 120 feet. Wilbur followed, covering 175 feet; then it was Orville's turn again, to fly 200 feet; and on the final flight Wilbur managed to cover 852 feet in 59 seconds. To be fair, Wilbur's distance through the air on the last flight was more than half-a-mile, but then and for many years afterwards the strength of the wind was certain to have a considerable effect on aircraft performance.

It was to be some time before the thought of establishing a speed record occurred to anyone, and longer still perhaps before the thought of official confirmation of such records was to come about. The latter did not occur until 1909. However, the framework for this came into existence soon after the Wright brothers' early flights. The responsible body for international aviation competitions, then as now, was the Fédération Aéronautique Internationale, which largely works through its affiliated national aero clubs. The Royal Aero Club was formed in 1901, while the Aéro Club de France is a little older.

Generally, the F.A.I. defines a record as the maximum performance obtained under the special conditions applicable for that record under the Sporting Code, and requires that recognition of an international record must be established under these conditions and certified by the F.A.I. The F.A.I. does recognize national records provided that the nationality of the record is that of the pilot and that a national aero club authorizes, controls and registers the attempt. It is up to the national aero club to apply to the F.A.I. for certification of a national record as a world record, and in the case of certain important spheres, such as speed, as an absolute world record. Notification to the F.A.I. of an attempt on a world or absolute record must be by telegram within two days of the completion of the attempt.

Not only are records classified regarding scope and aspect of performance, but the aircraft and apparatus are also classified for the F.A.I.'s purposes. Class A records are for free balloons, class B for dirigibles or airships, C for aeroplanes, seaplanes and amphibians, D for gliders, E for rotorcraft, F for model aircraft, G for parachutes, H for jetlift aircraft, I for man-powered aircraft, K for spacecraft and L for air cushion vehicles. An exception to the pilot's nationality rule is permitted in the case of model aircraft, in which the constructor's nationality is relevant, and if there is more than one crew member the captain's nationality is the deciding factor.

The F.A.I. recognizes speed records for piston-engined, turboprop and turbojet aircraft over distances of three kilometres with restricted altitude (a maximum of 100 metres or 328 ft); 15 to 25 kilometres without altitude restriction other than that level flight must be maintained within a tolerance of 100 metres altitude; and over closed circuits of 100 kilometres, 500 kilometres, 1,000 kilometres, 2,000 kilometres, 5,000 kilometres or 10,000 kilometres.

An aeroplane is defined by the F.A.I. as a 'heavier-than-air craft deriving its power from source, and which is sustained in the air by aerodynamic reaction on fixed surfaces during flight'.

In sporting regulations, the term 'landplane' is reserved for aircraft corresponding with the above definition and 'able to take-off from and alight only on ground,' while a seaplane or flying-boat also corresponds with these conditions except that it is 'a heavier-than-air craft able to take-off from and alight only on water', although, of course, an amphibian can take-off from and alight on either land or water.

Such conditions do not entirely bar recognition to an aircraft which is air-launched or launched with the aid of a catapult or ramp.

There is an element of endurance in all but the three-kilometre and 15/25-kilometre speed records, and for this reason these are the record attempts used in a definition of an absolute speed record. Certainly records over longer distances are slower.

The 15/25-kilometre length dates from 1955, whereas the three-kilometre rule dates from 1923. The early records were conducted over a variety of lengths, with a one-kilometre straight becoming standard after World War I, using an average timing of two runs. This became four runs in 1922. Up to 1927 a return run was required only over the three kilometres. The F.A.I. requires that the three kilometre course be a straight approved initially by the national aero club and extended at each extremity by an approach of at least 1,000 metres. A seaplane is restricted to an over-water course, but where there is difficulty in establishing a course over land for a landplane, an over-water course may be used. A maximum altitude of 500 metres throughout the record attempt is permitted, but during the approach to and the flights over the measured course, a maximum altitude of 100 metres must be observed. Four runs, two in each direction, must be made without landing and within a time lapse of 30 minutes, but more than four runs may be made with four consecutive runs being chosen for submission to the F.A.I. Only a 0.25 per cent margin of error in the determination of the speed is allowed, and the new record must be at least one per cent higher than the preceding record.

The differences between the requirements over the three-kilometre and the 15/25-kilometre course are matters largely of detail. Altitude for the latter is unlimited, but the tolerance accepted is confined to 100 metres, and the approach to the course must be of either 5,000 metres or 7,500 metres, with an altitude tolerance after take-off and before entry on to the approaches of 1,500 metres or 2,000 metres. The speed is the average of a return run over the course, without landing, and again within a time lapse of thirty minutes.

These are the basic rules. Some tolerance is allowed over the marking of the course, so long as it is approved. The pilot is not allowed to leave the aeroplane during the attempt, and if there is any accident during an attempt, everyone aboard must stay alive during the next 48 hours!

Chapter 1

THE EARLY RECORDS

The Wright brothers – first aviation meetings – rise of the
monoplane – Blériot – the Monocoque Deperdussin.

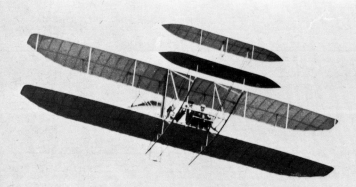

For several years following their breakthrough, the Wright brothers themselves made most of the technical advances in aviation. When it was still a notable achievement to have got an aeroplane into the air at all, their second model, the Flyer II, flew the first circle on 20 September 1904; their Flyer III became the first aeroplane to carry a passenger on 14 May 1908.

In the development of speed, they too set the pace. The Flyer I's modest speed of about 30 mph in 1903 was not exceeded by a European design until a French Farman-Voisin 1 biplane achieved about 40 mph late in 1908. The Wrights' fourth design, the Wright A, could rival the Farman-Voisin's speed while carrying a passenger, which was no mean achievement at a time when such a trivial matter as a lady passenger's skirt could upset the aerodynamic qualities of an aeroplane.

Nevertheless, the French were still the Wrights' nearest rivals during this period, and they continued to show considerable scepticism of the brothers' achievements, even going so far as to doubt the authenticity of the 1903 flights at Kitty Hawk. But the supremacy of the Americans was brought home to them on 8 August 1908, when Wilbur gave a short demonstration of the Wright A biplane at the Hunaudières race track near Le Mans. The French aeronauts immediately recognized that in both speed and manoeuvrability they were still far behind, and, to their credit, repentance for doubting the original flights was both genuine and public. Wilbur soon moved to the near-by military establishment of Camp d'Auvours, where two weeks later he established a number of height and endurance records, including one in which he flew over a circuit for 77 miles and won a prize of 20,000 francs from the Michelin concern.

While Wilbur was giving demonstrations in Europe (attracting so much attention that he decided to establish a flying school at Pau in the South of France in the hope of gaining a French army contract) and the Wright designs were being built under licence in Britain and France, Orville established a number of height records, including those of 200 and 310 feet, during military trials in the United States. By the end of the year Wilbur had raised the height record to 360 feet. These records emphasize the low flying altitudes of those days, which frequently required course alterations to avoid troops, cattle or even quite small buildings. The speeds then prevalent could result in aircraft being blown backwards by a strong headwind.

The first signs of a European contribution to the advancement of aviation came late in 1908, when Henry Farman, an Englishman who was later to take French nationality, flew a Farman-Voisin biplane on the first cross-country flights.

It was one of Wilbur's pupils at Pau, a Frenchman called Paul Tissandier, who set the world's first official air speed record, which (as already mentioned

The Wright A biplane of 1908, which not only demonstrated the superiority of the Americans to the French, but also established a world air speed record for one of Wilbur's pupils, Paul Tissandier, on 20 May 1909. The A was a development of the earlier Flyer III. It was flown extensively during trials in Europe and the United States, as well as being built by a number of licensees, and entered for the Rheims Aviation Meeting of 1909. (Photo: Science Museum, London.)

in Chapter 1) is a record ratified by the Fédération Aéronautique Internationale. Tissandier flew a Wright A biplane for $35\frac{3}{4}$ miles in 62 minutes on 20 May 1909, setting a speed record of 34.04 mph and becoming the first European to make a flight of more than one hour's duration.

But from this time on, the Wright brothers' lead in aviation design was to be overtaken by those who were able to draw upon their experience and achievement. Such is always the fate of the innovator.

The beginning of the end for the Wright brothers' supremacy and the over-shadowing of Tissandier's record came at the world's first aviation meeting, held on the Plain of Bethany just outside Rheims, from 22 to 29 August 1909. The Rheims aviation week (at which balloons and dirigibles were strangely much in evidence) was sponsored by the Champagne industry and officially known as the 'Grande Semaine d'Aviation de la Champagne'.

It is impossible to overestimate the significance of the Rheims meeting, where the first significant speed, height and endurance records were set, and European politicians and the public were brought face to face with aviation. Having set themselves up as the first patrons of this new art, the champagne industry offered generous prizes totalling 200,000 francs, thus providing a considerable incentive for the embryonic aircraft manufacturers, the best of whom were still at the 'cottage industry' stage.

Apart from the balloons and dirigibles, the then staggering total of thirty-eight aeroplanes was entered for the meeting, and of these no less than twenty-three flew, making a total of 120 take-offs and 84 flights of more than three miles. The fact that this all took place without a single fatality, and during a week of indifferent weather, with frequent interruptions to the flying pro-gramme because of high winds, made the achievement all the more impressive.

The most notable Rheims record was not one of speed, but of distance, with Henry Farman winning the Grand Prix with a flight of 180 kilometres in 3 hours 5 minutes, flying one of his own biplane designs, the Henry Farman III.

A succession of speed records came from Rheims, records being made only to be broken as the meeting progressed. The first of these came on 23 August, when the American, Glenn Curtiss, established a record of 43.35 mph, doing his best to brighten a generally disappointing day, with high winds preventing flying before 5 p.m. and a discouraging attendance. Earlier, Latham's Antoinette monoplane, according to *The Times*, 'took a header, striking the ground with such violence that the propeller was bent and the flying apparatus itself other-wise damaged.'

Although the record set up by Curtiss, flying one of his own designs, was short-lived, it is worth noting that it in turn had been an improvement over an unofficial performance by Lefebvre flying a Wright A the day before. The first American to fly after the Wrights in the United States, Curtiss was a fierce if not bitter rival of the brothers.

The next day, 24 August, again saw flying delayed until 5 p.m. while the aviators waited for the wind to drop. Hubert Latham was one of the first to fly, followed on to the circuit by Louis Blériot who, in one of his own monoplanes, 'flew grandly' to establish a record speed of 46.18 mph. Lefebvre followed in a Wright A with a performance which was daring but not quite fast enough. Blériot broke his own record by the end of the week, with 47.85 mph on 28 August, although earlier Curtiss had come close to challenging the Frenchman's supremacy while flying in the Gordon Bennett Cup Race.

The week was not without periods of friction, of which the most obvious must have been Hubert Latham's challenge of the award to Farman of the Grand Prix, worth £2,000. He alleged that Farman had disqualified himself by changing his aircraft's standard Vivinus water-cooled engine for an air-cooled Gnome rotary engine. The protest failed, and it is usually claimed that Latham was prompted to this action by the designer of his Antoinette, Léon Levavasseur. The irony of the situation was that the Antoinette's own powerplant, which suffered from a consistent lack of reliability, on more than one occasion landed Latham in water, literally if not deeply!

In spite of the weather and the friction, the latter understandable with so much at stake for the contestants, the first Rheims aviation week was an unqualified success. It enjoyed not only the patronage, but the attendance too, of the President of France, and the aviators performed their feats before a distinguished political and military audience.

Not least, the political figures included the then British Chancellor of the Exchequer, Lloyd George, who left Rheims stating that, 'Flying machines are no longer toys and dreams, they are an established fact. The possibilities of this new form of locomotion are infinite. I feel, as a Britisher, rather ashamed that we are so completely out of it.'

Long accustomed to an impregnable island status, the British had been rudely awakened by Blériot's successful cross-Channel flight on 25 July for a £1,000 prize offered by the *Daily Mail*. Lloyd George, as a member of the Cabinet, would also have been painfully aware of the decision by the Committee for Imperial Defence in February 1909 to support dirigible, rather than aeroplane development. During the period of the Rheims aviation week, the American-born, but by this time naturalized Briton, Samuel Cody, had managed to fly for seven miles at Aldershot. This was small consolation for Britain, bringing home, even more sharply, the poor state of British aviation.

The following year, 1910, saw a boom in aviation meetings, which were held throughout Europe, notably in France and Italy, in the United States, and even in Egypt. The first meeting in the British Isles was held at Doncaster, but not recognized by the Fédération Aéronautique Internationale, and, in any case, poorly attended. The first 'recognized' British meeting was at Blackpool, which still failed to compare with Rheims or Nice!

Nice had the distinction of being host to the first aviation meeting of 1910. It lasted from 10 to 25 April and offered prizes totalling 210,000 francs.

Hubert Latham won no less than 60,500 francs at Nice, no doubt compensating for his disappointments on the English Channel and at Rheims in the preceding year. He even succeeded in flying his Antoinette monoplane to a new world air speed record of 48.2 mph on 23 April, adding this to his 1909 altitude record of 1,486 feet. However, he did not escape his spot of engine failure and ditching in the sea later that day.

The leading competitor at Nice was one Efimoff, who flew a Farman biplane to win a total of 77,500 francs. This included a prize for a flight of more than three miles in length, which he managed with a speed of 35 mph. By comparison, the leading British competitor was the Hon. Charles Rolls, who won prizes totalling 6,000 francs while flying a Wright A biplane – probably the French-built aircraft which he bought to supplement a British-built machine of the same type.

Monoplanes continued to hold the lead at the second Rheims aviation week, held during July 1910. Weather conditions were much improved, and some commentators seemed almost disappointed with the lack of mishaps.

The weather remained excellent on 19 July, when Léon Morane flew a Blériot monoplane at a new record speed of 66.19 mph, winning a 10,000 francs prize. The record was established during a race in which the first three prizes went to Blériot machines, the second being flown by Alfred Leblanc and the third by a distance-record holder, Olieslager. Obviously, a spell of success for monoplanes in general at this stage, meant success for Blériot monoplanes in particular.

In much the same way that racing-car teams and drivers move across the world from one event to another, the early aviators turned out in force for the first major aviation meeting in the New World, held at Belmont Park, Long Island, New York, during late October.

The main event at Belmont Park was the Gordon Bennett International Speed Race for Aeroplanes, organized on an international team basis for the first time. Blériot machines predominated, with the Englishman Grahame White flying a Blériot to victory on 29 October. During the race Alfred Leblanc set a speed record of 68.2 mph in his 100-h.p. Blériot, but he was forced to retire from the event after his aircraft was caught by a gust of wind which loosened the fuel line to the motor. Ironically, the race was held on a clear, crisp day, with a moderate 10 mph breeze. It was probably a lack of altitude that accounted for the trouble.

Little public attention was accorded two further speed records by Leblanc, still flying a 100-h.p. Blériot monoplane. The new records came during 1911, with 69.48 mph on 12 April, and 77.68 mph on 12 June. The first was overshadowed by Prier's London to Paris flight, also in a Blériot, with Louis Blériot himself crediting Prier's success to the habit of flying high to avoid wind gusts near the ground. The other Leblanc record clashed with preparations for the Coronation of King George V in London, the first occasion for the banning of aircraft flights over London and Windsor. Could it have been a hint to the intrepid aviators as to just how much confidence the Board of Trade had in them and their machines!

Fitting neatly into the gap between Leblanc's second and third speed records, another Frenchman, Edouard Nieuport, flew one of his own monoplanes to a record of 74.42 mph at Mourmela on 11 May 1911. This flight was hailed as being 'remarkable', and so it must have been, with an engine of only 28 h.p. and two cylinders! A ten-kilometre course was covered in a fraction over five minutes by the Nieuport monoplane.

In case posterity should doubt his efforts, Nieuport broke his first record with one of 80.82 mph on 16 June 1911, at the Camp de Châlons, breaking this in turn with a speed of 82.73 mph at the same place only five days later. Oddly enough, his achievements were marked by something less than frantic excitement. But tastes of news editors, if not also their readers, are nothing if not fickle.

However, the Antoinettes, Blériots and Nieuports which had put the Wright brothers' machines so firmly in their place were to be superseded in their turn by a new aeroplane, the so-called Monocoque Deperdussin.

The monocoque concept was invented by a Swiss, Ruchonnet, and as such has probably been Switzerland's sole contribution to aviation technology, although there can be little doubt as to the ability of the Swiss to use the results of other people's work for military and civil aviation. It is an idealized concept of a completely hollow structure or shell, meaning that the shell of the airframe actually bears the load or stress rather than just acting as the cover for a clumsy frame. The monocoque concept is not exclusive to aviation; today it applies to most motor-cars, which depend on the body for strength, thus eliminating the need for a chassis. A good monocoque structure is stronger than a conventional structure.

Designed by Bechereau, the Monocoque Deperdussin used a wooden fuselage. It was a very modern aeroplane in appearance, with a streamlined monoplane configuration and the engine, which on the early models was a 140-h.p. Gnome rotary, was partially enclosed. This made the rival aeroplanes, including even the Blériots, appear dated by comparison. The Monocoque Deperdussin can probably be regarded as the first aeroplane in history to be designed largely for racing and record-breaking, with little operational requirement. Viewed in this light, the design had more in common with the record-breakers of the post-World War I period than with those of the early days of aviation.

A succession of speed records fell to the Monocoque Deperdussin, starting at Pau on 13 January 1912, when Jules Védrines flew the aircraft to a speed of 90.20 mph over a five-kilometre circuit. This was only a start, for on 22 February the aircraft scraped past the magical 100 mph mark with a speed of 100.23 mph, again with Védrines as pilot. All the man in the street heard of this epoch-making achievement was a terse dispatch from Reuters; yet the record was the highlight of a day in which the aircraft covered 100 kilometres in 37 minutes and 200 kilometres in 75 minutes.

The Monocoque Deperdussin and Védrines then settled down to inching the speed record steadily upwards, after the earlier strides in which the aeroplane, with a reproduction of Leonardo da Vinci's 'Mona Lisa' on its fuselage, had demonstrated its superiority over its competitors.

A marginal increase in the official record was achieved on 29 February with 100.95 mph at Pau, which by this time had become something of a Mecca for speed-conscious aviators and their not always so very trusty steeds. Still at Pau, Védrines pushed the record higher, to 103.66 mph on 1 March, and then to 104.34 mph on 2 March. Lacking the excitement of a race or the sparkle of an aviation meeting, these records attracted little interest at the time.

Speed alone is not always useful, and for this reason the two-hour speed record of 147 miles established on 1 March 1912, by another Frenchman, Tabuteau, also at Pau, but flying a 50-h.p. Gnome-powered Morane-Saulnier, is worthy of mention. The Morane-Saulnier, a make which was to become well known in later years, demonstrated not just speed with distance, or endurance, but reliability. Engine reliability was still not all that it could be, and was the weakness of many designs.

The 1912 Gordon Bennett races were due to be held at Chicago, a result of the American victory two years earlier. Qualifying trials were held at Rheims during July, in order to select a French team of three aircraft. The trials gave the Védrines-Deperdussin combination the opportunity of setting yet another record of 106.12 mph on 13 July. The following day Armand Deperdussin, the owner of the plane's builders, was made a Chevalier of the Legion of Honour in recognition of his services to French aviation.

The French could well feel pleased with themselves, for they had overtaken the Americans and were now far ahead of any other country in aviation development. Not until World War I awakened Britain to the strategic possibilities of aviation did she overtake France. It was longer still before the Americans were to regain their old supremacy.

Needless to say, Védrines in the Monocoque Deperdussin not only won the 1912 Gordon Bennett Cup race at Chicago, but in so doing established a new official record of 108.18 mph on 9 September. He managed to resist the strong temptation to raise the record further during what was left of the year. A reminder that the Deperdussin was merely a speed machine comes with the knowledge that the height record for 1912, 18,405 feet, was held by a Morane-Saulnier, and the endurance record, 628 miles, went to a Maurice Farman which had flown over a circuit, rather than across country.

In aviation, as in everything else, no forward-looking designer can afford to sit still, and even an obviously advanced and successful design must be improved if it is to stay ahead of rivals and imitators. The fact that a fast production scout biplane, such as the Sopwith Tabloid, could only manage 92 mph against the Monocoque Deperdussin's 108 mph, was no excuse for complacency. Necessity dictated that the Deperdussin should greet 1913 with a new engine, a 160-h.p. Gnome, and a new pilot, Maurice Prévost.

The 1913 Gordon Bennett races were held during the Rheims aviation week, which had become firmly established as an annual event. After establishing a new record, the first for the up-rated aircraft, of 111.74 mph on 17 June, Prévost appeared at Rheims during September to set a new record of 119.25 mph on the 27th, and to win the Gordon Bennett Cup on 29 September with a speed of 126.67 mph, flying in a dead calm. This was the last of the Monocoque Deperdussin speed records, but it should be remembered that this one basic aircraft type moved mankind some 50 per cent faster than any of its predecessors, raising the speed record from 82 mph to 126 mph.

If this seemed to be no mean feat, it was also not the complete story. There were no British or American competitors in the 1913 Cup race, and of the four finalists, three were Deperdussins. The second competitor, Emile Védrines, brother of Jules, flew the only non-Deperdussin entry, a 160-h.p. Gnome-powered Pinnier monoplane. Third came one Gizellet, who had flown his Monocoque Deperdussin so high, 17,868 feet and within 2,000 feet of the then altitude record, that he suffered terribly from cold and needed oxygen.

That the Pinnier came second shows that others were ready to match the performance even of an aircraft as far ahead as the Monocoque Deperdussin.

Rheims in 1913 must have been exciting, and one commentator was even moved to remark that the large attendance was 'quite like old times'. One competitor so forgot himself as to break the rules and fly over the spectators, earning himself a fine of £8 – a penalty applauded as setting a good example for aviation meeting organizers elsewhere, particularly in the United Kingdom.

The monoplane, which had predominated so heavily at Rheims in 1913, was shortly to be put in its place, somewhat sadly and unfairly for all those who had devoted so much effort and expense towards its development.

Blériot himself discovered a weakness in one of his military production designs and in a typically responsible manner drew the attention of his customers, the British and French Governments, to the defect while he hastily put modifications in hand. Nevertheless, the aircraft were grounded and a strong anti-monoplane bias became apparent in the purchasing policies of both Governments. Thus the biplane, with some triplane assistance, had the Allied field largely to itself during World War I. Such was the price for Blériot's honesty and integrity!

The absolute speed records and the Gordon Bennett races do not present a complete picture of 1913 and 1914 with respect to speed in the air. Normally a landplane, the Monocoque Deperdussin design appeared in float-plane form for the first Schneider Air Trophy Contest for Hydro Aeroplanes in April 1913. The aircraft won, but with an average speed of only 45.75 mph because of the need to make an extra lap due to a dispute over crossing the finishing line. In 1914 a float-plane version of the Sopwith Tabloid won with a maximum speed of 86.78 mph, foretelling the re-emergence of the biplane and the emergence of the British aircraft industry as a power to be reckoned with. However, the Schneider Trophy races deserve a chapter to themselves and are dealt with accordingly.

Opposite above, left: Wilbur Wright (1867–1912), the elder brother and the leader. It was Wilbur who brought the Wright A to Europe and stayed on to found a flying school after giving demonstrations in the hope of obtaining orders from the French military authorities. To Wilbur must go the credit for taking the initiative in the aeronautical research which led directly to the first powered aeroplane flight in history. *Right:* Orville Wright (1871–1948), who remained in the United States after Wilbur left for Europe. Orville had made the first heavier-than-air powered flight, and on 9 September 1908 he made the first flight of more than an hour. (Photos: Science Museum, London.)

Below right: Louis Blériot flying across the English Channel in one of his monoplanes early in the morning of 25 July 1909. Although the Blériot machines were not without success in speed events, and Blériot himself, and later his fellow countrymen, Morane and Leblanc, established speed records in Blériot monoplanes, it is for achievements such as the cross-Channel flights that they are best remembered. (Photo: Science Museum, London.)

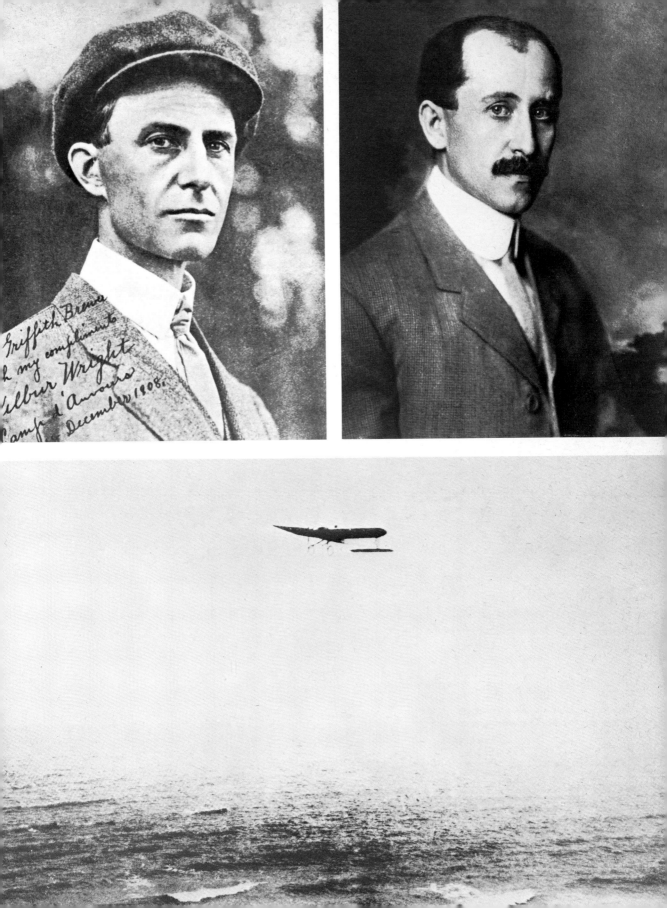

Opposite: Anyone with a spark of life or romance in him should have been moved by this effective advertisement for the first aviation meeting in 1910, at Nice in April. The Antoinette with its Anzani engine may owe much to the imagination of the artist, but it has a larger-than-life quality.

Top: Hubert Latham in the cockpit of an Antoinette monoplane at the Rheims Aviation Meeting. A singularly unfortunate aviator, his flights were often punctuated by a ditching in the sea, not least during his unsuccessful attempt to be first across the English Channel. (Photo: Science Museum, London.)

Above: An Antoinette monoplane being flown by Latham in England at Brooklands, near Woking. At this time, and between the wars, Brooklands was more famous as a motor racing circuit, although today it is also the home of an aircraft factory building sub-assemblies for the Concorde supersonic airliner and the One-Eleven short-haul jet airliner. (Photo: Science Museum, London.)

Right: A rather less sparkling affair was the first British aviation meeting in October 1909 at Doncaster. Lacking official recognition and a suitable sponsor, the meeting was a disappointment, not least to the Great Northern Railway which obviously saw traffic potential in the occasion.

MEETING D'AVIATION
NICE
10-25
AVRIL
1910

CH. BSOR

WORLD WAR I

Official records suspended, but many of them broken – the fighter.

An aeroplane does not always have to be fast to be effective, even if it is supposed to be a fighter! This is a full-sized flying replica of the Vickers F.2B Gunbus, which did much to counter the so-called 'Fokker Scourge' of World War I, although having a fairly unexciting top speed itself. (Photo: British Aircraft Corporation.)

The outbreak of war in 1914 is popularly presented as having been the making of the aeroplane. For most it is almost an article of faith that aviation advanced more in four years of conflict than could have been possible in twenty years of peace. An opposing view, presented by Sir Richard Fairey, is that war hinders rather than assists aeronautical development. To accept either argument at its face value is to accept an over-simplification and to ignore the facts.

Obviously official air speed records came to an end with the start of World War I, for reasons of secrecy and for want of the sporting events with which many record-breaking attempts had been associated. Nevertheless, performance did improve, with the 1913 record of 126.67 mph rising to the first official record of the post-war period, in 1920, of 171.05 mph. The pre-war record speeds became commonplace during the war years.

Yet, all this progress was only the result of stretching existing technology to the full. The war years were a desert if one looked for new technology. While the achievements of the immediate pre-war period gained a wider and earlier acceptance than would have been given them in peacetime, there was relatively little original research unless one includes that of Professor Hugo Junkers in Germany. On the outbreak of war, Sikorsky had already built his first multi-engined types, and the record-breaking and racing aircraft were set to become the first scouts.

But the fastest aircraft of the years immediately preceding the war did not take part in any battle. The Monocoque Deperdussin was largely a racing design, partly because of the break-up of the Deperdussin concern after the arrest of Armand Deperdussin on fraud charges in August, 1913, and partly because of the distrust of the monoplane concept in the minds of British and French military authorities.

This is not to say that all military aircraft in Allied service in 1914 were biplanes. Nieuport and Morane-Saulnier monoplanes were used side by side with Deperdussin monoplanes, although not of monocoque construction. The Germans did not share the distrust of the monoplane. Their standard aircraft, the Taube (dove) and the Fokker E.I, were both monoplanes.

Blériot acquired the Deperdussin works and, employing Bechereau as chief designer, established S.P.A.D. (Societé pour l'Aviation et ses Derives), building biplanes!

The fastest aircraft in Allied service at the outbreak of war were the Sopwith Tabloid and the Bristol Scout with maximum speeds slightly in excess of 100 mph. Less illustrious types, such as the Royal Aircraft Factory B.E.2A, could only reach 75 mph, and the Avro 504 was soon relegated to training duties, in which it excelled. The Nieuport and Morane-Saulnier monoplanes had a performance comparable to that of the Bristol and the Sopwith, and the Fokker would not have been much slower. The curiously named Taube did not excel at speed, and performance varied widely depending upon the extent to which the original Rumpler-Etrich design was varied by the many manufacturers licensed to build the type in Germany.

Not every new aircraft type which followed during the war added an increase in speed. The bombers were slower, as befits aircraft designed to carry a load, but, strangely, so were fighters such as the Airco D.H.2 and the Vickers Gunbus. These were pusher propeller types built to counter the so-called 'Fokker Scourge' of propeller synchronized machine gun-fitted D.IIIs. The D.H.2 could manage a sluggish 80 mph, and the Gunbus was even more staid at 70 mph – but both aircraft were effective.

Naturally, comparisons between speed machines, specially tuned and stripped of unnecessary weight, and warplanes, carrying adequate fuel, ammunition, and sometimes a second man as well, are approximate. At first, however, the aircraft were unarmed, and then a rifle or just a revolver would be provided. It was some little time before machine guns were fitted and the fastest of the aircraft, the various scouts and the Tabloid, became fighting-scouts, or the forerunners of the fighter.

Developments of the Tabloid were not long in coming. First of these was the Pup of 1916, variously powered with 80-h.p. Gnome or Le Rhône rotary engines, or with a 100-h.p. Gnome Monosoupape with which a speed of 110 mph was attainable. A later Sopwith of note was the Camel, with a 130-h.p. Clerget engine giving a maximum speed of 120 mph; this aircraft entered service with the Royal Flying Corps and the Royal Naval Air Service towards the end of 1916. Later still, the Sopwith Dolphin could achieve 130 mph.

Camels were used by many British aces. They gained at least some of their fame from shipboard and barge take-offs and not a little notoriety for being bad mounts for the careless or inexperienced pilot. Perhaps these biplanes had something of their quadruped namesake's uncertain temper!

It was a Camel flown by Captain Roy Brown, R.F.C., which shot down Baron Manfred von Richthofen in a Fokker Dr.1 in April 1918.

The Royal Aircraft Factory S.E.5a, with a 200-h.p. Hispano-Suiza engine, was capable of attaining 130 mph near sea level. It was the mount of many British aces, including the then leading fighter pilots of World War I: Majors Micky Mannock and James McCudden, R.F.C., and Captain Albert Ball, R.F.C. Ball also used a Nieuport 17, which he is known to have preferred, and a number of French aces also used this machine.

Other notable Allied aircraft were the Italian Ansaldo, capable of 140 mph from its six-in-line engine; it represented not only the rapid rise of Italy's aircraft industry during the war, but foretold, in some measure, her post-war achievements. The S.P.A.D. XIII was a favourite of American pilots, deprived of any national warplane by the then poor state of the American industry, which was largely occupied with licence-building British and French designs, including the Airco D.H.4 bomber.

The German equivalent of these aircraft was the Fokker D.VII, of which Hauptmann Rudolf Berthold, with 44 confirmed victories, was a brilliant pilot. The German Albatros concern, which had built aircraft in which such men as Boelcke and Immelman flew to fame, had by this time lost much of its early momentum.

A typical late war fighter, intended to soldier on throughout the 1920s, was the Bristol F.2B Fighter biplane, with a maximum speed of only 125 mph. The 'Brisfit' was a rugged machine, however, and more in the fighter-bomber mould, well suited to the R.A.F.'s post-war police duties.

Speed is not everything for a warplane, but nevertheless, speeds had risen from well below the 100 mph mark for the fastest fighters to 140 mph, and altitudes had soared to as much as 21,000 feet for many aircraft. The bombers too were ready to pave the way for the first airliners, with speeds of around 100 mph for a D.H.4a, capable of carrying a 450-lb. bomb load and possessing an endurance of four hours.

The rotary engine reached the peak of its development, and designers were forced to study the in-line and radial engines with renewed interest, in more than one instance hastily 'hotting-up' a car engine to obtain a suitable in-line powerplant. In short, there had been progress, not of a dramatic or forward-looking kind, but of the type which squeezes the last ounce of value out of what is available at the time. This was the meaning of World War I for the aeroplane, and the high performance aeroplane in particular.

Most significant, America had been spurred back into the race, and Italy had awakened to the possibilities of aeronautics. In the years to come, these were the developments which would carry much weight, politically and economically as well as technically.

A model of the German Rumpler Taube monoplane, which formed the mainstay of the Military Aviation Service's frontline strength at the outbreak of World War I in 1914. The Taube was available in a variety of forms, including a rudimentary cabin monoplane and also a metal version known appropriately as the 'Stahltaube'. It was built by most German aircraft manufacturers in order to supplement Rumpler's own production. The extensive bracing of the early monoplane is nowhere more in evidence than on the Taube, and the drag generated by this outweighed the aerodynamic advantages over the biplane. (Crown Copyright, Science Museum, London.)

Left: A nine-cylinder version of the Gnome Monosoupape rotary engine, which powered many aircraft before and during the early years of World War I, including the Sopwith Tabloid. This preserved example was built in England. For the pilot the rotary engine meant a pungent smell of castor oil, and for the designer it meant a limitation because of the instability generated as power outputs increased. The rotary engine, based on a concept formulated by the Australian pioneer, Hargrave, was a significant step forward. (Crown Copyright, Science Museum, London.)

The 'Fokker Scourge' came with this Fokker E.111 Eindecker monoplane, with an Oberursal rotary engine and a propeller-synchronized machine gun, which enabled it to wreak havoc amongst the Allies until checked by the Gunbus and D.H.2 pusher biplanes. (Photo: Richard E. Gardner.)

The early years of aviation were dominated by the landplane. Not until 1910 did the float-plane, or hydro-aeroplane, first appear in powered form, with a tentative and courageous flight by the Frenchman, Henri Fabre. In 1911 the American, Glenn Curtiss, built and successfully tested the first practical float-plane, after which a limited number of float-planes and flying-boats began to appear, although these were less satisfactory in performance than their wheeled counterparts.

This relative neglect of the hydro-aeroplane disappointed the young, wealthy Frenchman, Jacques Schneider, who was convinced that the key to the successful development of the aeroplane lay with the float-plane and the flying-boat.

The son of an armaments manufacturer, Schneider was a mining engineer by profession, although his interest in aeronautics had been kindled by the demonstration flights of Wilbur Wright in France during 1908. He gained his aeroplane and aerostat pilots' qualifications during 1910, and in 1913 established a French altitude record of more than 30,000 feet in his balloon *Icare*, although an old injury prevented him from taking a place in the forefront of the pioneering pilots.

Nevertheless, Schneider's contribution to the development of the hydro-aeroplane – in effect the seaplane although the term had still to be coined – because of the speed limitations of the flying-boat, was to be immense. At a banquet held on 5 December 1912, after that year's Gordon Bennett air race for landplanes, he presented a trophy, the Jacques Schneider Trophy for Hydro-Aeroplanes, to be contested for annually by national teams of hydro-aeroplanes. In common with the Gordon Bennet Cup, the Schneider Trophy contest was to be under the overall supervision of the Fédération Aéronautique Internationale, with competitors entered by national air clubs affiliated with the F.A.I. Qualifying tests were to be held in order to find a team of three from each participating country, and the winning country of each year would be responsible for holding the following year's contest. Three victories by any one country in five years would qualify for an outright award of the Schneider Trophy.

Hardly an object of beauty, the Schneider Trophy consists of a base of marble decorated with bronze, upon which a silver nude winged figure is depicted kissing a zephyr recumbent in the crest of a wave. This wave contains two other zephyrs and the head of the sea god, Neptune, all in silver. Many who see the Trophy for the first time immediately assume that the winged figure must be Icarus, but this is to ignore the ample evidence of femininity!

The first Schneider Trophy contest bore little resemblance to the major event of the post-World War I period; it was included, almost as an afterthought, in the Monaco Hydro-Aeroplane Meeting of 3 to 17 April 1913, organized by the International Sporting Club of Monaco. On the other hand, the Trophy contest and its eliminating trial – somewhat unnecessary because the number of would-be French contestants had been whittled down to three by the weather, inexperienced flying and rudimentary aircraft design, leaving only one other competitor – were the only really successful parts of the whole meeting.

Most of the Monaco meeting was concerned with the preliminary trials for the main event, the hydro-aeroplane Grand Prix de Monaco, but out of sixteen competitors, only seven aircraft survived for the start. Miraculously, there was only one fatal accident. Only two of the three starters survived the race – by seeking shelter as the race was finally abandoned.

Not for the last time, the four aircraft competing in the first Schneider Trophy contest were conversions of landplanes. The Frenchman Maurice Prévost had a Monocoque Deperdussin, specially fitted with floats for the occasion. The almost equally famous French stuntman and long-distance flight pioneer, Roland Garros, had a Morane-Saulnier, while the third member of the French team, Dr Gabriel Espanet, and their American rival, Charles Weyman, both used Nieuports. These aircraft were all monoplanes. Their task was to cover the ten-kilometre Grand Prix course running between Monte Carlo and Cape St Martin twenty-eight times, with the first half-lap spent taxying on the surface to prove that the aircraft were true hydro-aeroplanes and not freak machines.

This was to be no mean feat. Float design was primitive, although Weyman's Nieuport had stepped floats based upon hydro-planing experience. Even if the take-off could be tackled successfully, once in the air the floats and the tangled web of supporting struts and wires were not conducive to good aerodynamics. Fortunately, the competitors had the only good weather of the fortnight. With just a gentle breeze in place of the devastating mistral, they decided upon an early start, at 8 a.m.

Prévost took off first, followed by Garros. Garros literally drowned his engine in spray and had to return to the slipway. Finally came Espanet and Weyman.

Engine trouble in the eighth lap forced Espanet to retire; then Prévost, who had misread the rules, taxied over the finishing line instead of flying over. Procrastination stole much of Prévost's time, for he refused to make a further final lap in the face of seemingly certain defeat by Weyman, only to find the American forced to land with an oil leak in the twenty-fifth lap. After Weyman's trouble, Prévost relented and flew once more around the course, but since his time for the race had run without stopping from his original start, the delay had reduced his winning average speed from 61 mph to 45 mph.

Garros, who had abandoned a second attempt while the contest was in full cry, returned to the start line for a third time and eventually finished in second place, although his speeds of up to 60 mph were of little use when averaged together with the time wasted on the ground in repairs and indecision.

The Aéro Club de France was left to stage the next Schneider Trophy contest, and in fact they returned to the Monaco course of the previous year, holding the event on Saturday, 18 April. The 1914 contest was very much more in line with Jacques Schneider's main objective, with a truly international competition and a marked improvement in performances. It also pointed towards the shape of the aeroplane over the next few years and foretold the ending of French domination of high performance aircraft development.

An 8 a.m. start was decided, and once again the competitors were blessed with good weather and a light breeze. The main departure from the 1913 contest was that the half-lap taxi run was replaced by two touchdowns, of no particular length, during the first lap. So much for seaworthiness!

The period immediately preceding the day of the contest had been hilarious for the French. These masters of the air had adopted a very superior attitude, which was in no way diminished by their suffering a number of surprises. Maurice Prévost's Monocoque Deperdussin failed to become airborne in a choppy sea during the eliminating tests, thus denying the French their most experienced pilot and most promising aircraft. The Deperdussin had had its 160-h.p. engine of the preceding year replaced by a 200-h.p. unit. Still, the French remained confident that their team, with Gabriel Espanet and Pierre Levasseur both in 160-h.p. Nieuports and Roland Garros in a 160-h.p. Morane-Saulnier, would be more than a match for the other entrants.

Only Charles Weyman, still with a Nieuport, was taken seriously by the French. The German Aviatik Arrow, flown by Ernst Stoeffler, was dismissed out of hand as being too slow, so was the Swiss F.B.A. flying-boat with a 100 h.p. Gnome Monosoupape. Little attention was paid to one of the British entries, a Morane-Saulnier flown by Lord Carbery, and the other, a 100-h.p. Gnome Monosoupape-powered Sopwith Tabloid float-plane of remarkably small dimensions, was treated with amusement.

Outright ridicule was reserved for the other American contestant, William Thaw, who had entered a battered Curtiss flying-boat. He was eventually shamed into borrowing a 160-h.p. Monocoque Deperdussin.

In the event, the Aviatik was a non-starter, as would Carbery have been had he not been able to borrow a Monocoque Deperdussin.

The starting order was Levasseur, Espanet, Burri in the F.B.A., then the Englishman Howard Pixton in the Sopwith Tabloid, putting on such a sparkling performance from the start that Weyman and Garros decided that participation in the contest would be hopeless. A spell of engine trouble during the middle of the contest reduced the Tabloid's lead slightly, but the small machine soon recovered and impressed the hitherto cynical French with its speed and manoeuvrability. Carbery ran into difficulties during the alighting tests and wrecked the engine of his machine during a second start. Burri ran out of fuel on the twenty-third lap.

An average speed of no less than 86.78 mph was recorded by the Tabloid during the contest, and as an encore the aircraft remained airborne to establish a hydro-aeroplane record of 86.7 mph for the 300 kilometres, then landing in a sea which had suddenly become rough as if to prove that this particular design was not a freak.

Naturally, the Schneider Trophy contest did not take place during World War I, and the right, or more accurately the obligation, to stage the first competition of the post-war period belonged to the Royal Aero Club of the United Kingdom. The 1919 contest, held on 10 September, is on record as one of the worst organized, if not the worst. The whole affair collapsed in a comedy – but fortunately not a tragedy – of errors.

The first mistake was to hold the contest at Bournemouth, a resort which in spite of other attractions, completely lacked facilities for a seaplane meeting. Competitors had to fly from their base at Cowes to Bournemouth, where the last minute preparations were put in hand on the morning of the race, 10 September 1919, amidst a throng of holiday-makers, on the beach and in the water. A second mistake was to pay insufficient attention to the weather, and finally there was added confusion over the course and the starting procedure.

Ten laps were to be made of a twenty-nautical-mile course, running from Bournemouth to Swanage, then to Christchurch and back to Bournemouth. Two alightings were to be made during the first lap.

A complete team was fielded by Great Britain for the first time, with Vincent Nichol flying a Fairey IIIA, Harry Hawker a Sopwith seaplane and Basil Hobbs a Supermarine Sea Lion flying-boat. The French team consisted of a Nieuport and a Spad, and the first Italian Schneider contest entry was a Savoia S.13 flying-boat, flown by Sergeant Guido Jannello.

The original intention had been to start the race at 2.30 p.m., but the mist of early morning had returned by 2 p.m., causing a postponement to 6 p.m. This was welcomed by the French who had sustained damage to their floats on landing after the flight from Cowes. Further changes of plan brought the starting time forward to 4.30 p.m. and then pushed it back slightly to 4.45 p.m. At this the French decided to withdraw from the event.

The three British entrants all took off within a short period, in spite of confusion over the starting arrangements, and flew into dense fog at Swanage.

As a result, the Fairey and the Sopwith abandoned the contest on returning to Bournemouth at the end of the first lap. The Sea Lion landed in Swanage Bay to find its bearings, sustained damage on taking off, and promptly sank on landing at Bournemouth!

Meanwhile, Jannello had taken off after starting correctly, and a decision over whether to abandon the contest was deferred while the Royal Aero Club officials waited to see whether or not he could complete the course. Jannello, however, appeared to be flying superbly, except for the final lap on which his time seemed to be exceptionally short. In order to avoid any risk of disqualification and for fear that he may have missed a marker boat (he had in fact missed that at Christchurch), he was sent round for a further lap, but he ran out of fuel and had to make an emergency landing. It was at this time that the organizers discovered that Jannello had been flying round the reserve marker boat moored in Studland Bay instead of the Swanage marker, and he was disqualified.

The mistake was an easy one to make, and the race was declared void despite Italian protests. An attempt by the Royal Aero Club to relent and award Italy the Trophy was over-ruled by the F.A.I., although the Italians were allowed, as a consolation prize, to stage the next contest.

Although an improvement on 1919, the 1920 contest was far from being a success.

It was not possible to send a British team to Venice for the 1920 contest, because of a shortage of funds and the lack of a suitable aircraft. A French entry failed to materialize. In part, this state of affairs may have been due to a change in the rules requiring all competitors to carry 300 kilograms of ballast. This requirement, the result of Italian pressure, was dropped for 1921.

The Italians were lucky to be able to compete themselves, in spite of the absence of a foreign team. Weather and still inadequate design put two of Italy's three aircraft out of the competition. Not for the last time, one aircraft, a Savoia S.12 flying-boat with Lieutenant Luigi Bologna of the Marina Militare Italiana, remained to fly ten laps of 37.2 kilometres each on 20 September. Determined not to falter during the contest, Bologna did not attempt to push the pace and his average speed was only 107.22 mph.

It is not surprising that many commentators were becoming cynical about an annual non-event of such disappointing proportions. The Italian attempt to win the Trophy outright was almost fanatical, with no less than sixteen aircraft entering the eliminating tests for the 1921 team.

For 1921 the event would consist of sixteen laps of 24.6 kilometres each, with alighting tests during the first lap. Venice was again to be the venue, and the date was fixed for 11 August.

The Italian team consisted of two Macchi M.7 flying-boats, flown by Giovanni de Briganti and Piero Corgnolino, and a Macchi M.19 flying-boat, to be flown by Arturo Zanetti. France entered Sadi Lecointe in a Nieuport-Delage seaplane. In the event, the float supports of the Nieuport-Delage collapsed during the alighting tests, while Zanetti's M.19 managed 141 mph on its first lap before catching fire and retiring, and Corgnolino's M.7 ran out of fuel on the last lap. Out of this debâcle Briganti's M.7 soldiered on at a fairly sedate 117.9 mph.

Fortunately, an event more worthy of the vision of Jacques Schneider was forthcoming at Naples in 1922 with thirteen laps of 28.5 kilometres each. The French and Italian teams were given government assistance for the first time. This, however, failed to prevent the French C.A.M.S.36 flying-boats from being eliminated during the preliminary tests, leaving the Italians to face a purely private-enterprise British rival.

This sole British entrant was the Supermarine Sea Lion II flying-boat, flown by Henry Biard. It had been re-designed by the young Reginald Mitchell, on the basis of the Sea Lion of the 1919 competition. Biard faced an experienced Italian team comprising Alessandro Passaleva in a Savoia S.51, Arturo Zanetti in a Macchi M.17 and Piero Corgnolino in a Macchi M.7. Like the Sea Lion, these aircraft were flying-boats.

Unluckily for the Italians, their best hope, the S.51, was damaged before the start of the contest, and although able to compete, it did so at a distinct disadvantage.

The event was timed to start at 4 p.m. on Saturday, 12 August. The heat of an exceptionally fine day necessitated this late start.

Almost from the word go, the Sea Lion II showed that, for the first time in the post-war period, here was to be a victor worthy of the Trophy, impressing the Italians not so much by its speed, but by its exceptional manoeuvrability which enabled Biard to keep close to the pylons at the ends of the course. All four aircraft completed the course, with the Sea Lion II's winning average being a close 145.7 mph against the S.51's 143.5 mph. The S.51 would probably have won the contest had it not been for its misfortunes beforehand, for the following December it established a flying-boat record of 174.08 mph.

It was the 1923 contest at Cowes on the Isle of Wight that finally achieved the calibre of competition which Jacques Schneider always had in mind, surpassing even the 1914 event. The Royal Aero Club had learnt the lessons of Bournemouth.

The contest was fixed for 28 September. The navigability tests would be held on the preceding day. The course would be five laps of 37.2 nautical miles each.

Originally Britain, France, Italy and the United States all intended to field teams, but a lack of support by the Italian Government meant a last minute withdrawal by their team. The British entry was cut to a single Supermarine Sea Lion III flying-boat, again to be flown by Henry Biard, by the failure of the other aircraft, the Blackburn Pellet, during the navigability tests. The French C.A.M.S.38 flying-boat of Lieutenant de Vaisseau Havel was also the only entry left from that country after the C.A.M.S.36 and Latecoere L.1 flying-boats failed to start. Least depleted was the United States entry, but even their team suffered the withdrawal of the Navy-built TR-3 float-plane before the start of the event. This left two U.S. Navy Curtiss CR-3 seaplanes to be flown by Lieutenants David Rittenhouse and Rutledge Irvine, U.S.N.

In excellent weather Rittenhouse won at an average speed of 177.38 mph, the other CR-3 came a fairly close second with 173.46 mph, and the Sea Lion trailed behind at 157.17 mph. At least the British entry finished, unlike the C.A.M.S.38, which retired during the first lap while flying at the eastern end of the course near Selsey. It was obvious that the Europeans had nothing to compare with the CR-3 biplanes, with their 465-h.p. Curtiss D-12 engines, although the Sea Lion's 550-h.p. Napier Lion was obviously hampered by being placed in a flying-boat instead of in the faster seaplane.

A variety of circumstances prevented the European nations from sending teams to Baltimore in an attempt to regain the Trophy in 1924, and the Americans could legitimately have run over the course themselves, as the Italians had done in 1920, and won the Trophy for a second time. Instead, the National Aeronautical Association, which was the responsible body for aviation competitions in the United States, made the offer to the Royal Aero Club, which had been the last to withdraw from the event, that the 1924 contest be cancelled. Needless to say, this was gratefully accepted.

Baltimore was again the venue for the 1925 contest, also organized by the National Aeronautical Association. The course, in Chesapeake Bay, was to be completed in seven laps of 50 kilometres each. The preliminary trials were to be followed by a six-hour mooring-out period before the contest, in an attempt to eliminate unseaworthy, and therefore freak, machines. Originally scheduled for 24 October, the actual event was postponed until the 27th because of adverse weather which had persisted throughout the preparations for the contest.

The American team consisted of Lieutenant James Doolittle of the U.S. Army, Lieutenants George Cuddihy and Ralph Ofstie, U.S. Navy, all flying Curtiss R3C-2 seaplanes with 600-h.p. Curtiss V-1400 engines. Italy had entered two Macchi M.33 flying-boats, to be flown by Giovanni de Briganti and Riccardo Morselli; Great Britain had entered Henry Biard with a Supermarine S.4 cantilever monoplane seaplane, and Bert Hinkler and Herbert Broad with Gloster III biplane seaplanes. Engine trouble caused the withdrawal of Morselli's Macchi, the S.4 crashed during trials because of wing flutter and Hinkler's Gloster III was damaged by severe wave buffeting while landing for the navigability trials.

The end result was simply a postponement of what could have happened a year earlier. Although Ofstie was forced to retire during the sixth lap and Cuddihy followed suit in the final lap, Doolittle raced home with an average speed of 232.57 mph, leaving Broad's Gloster trailing behind with a slow 199.17 mph and Briganti's Macchi still farther back with a hopeless 168.44 mph.

Still denied Government support, even though test aircraft bought by the Air Ministry could be loaned back to the manufacturers for the Schneider Trophy contest, the British team stayed at home in 1926 to lick its wounds. British hopes that the contest would be postponed once again were dashed by Italian entry, although eventually the F.A.I. ruled that after 1927 the contest would become biennial.

In 1926 the Americans were faced with staging the contest once again. They chose Hampton Roads, Virginia, and a 50-kilometre, seven-lap course into Chesapeake Bay and past Newport News to the naval base at Norfolk. Inspired or pressurized by Benito Mussolini, Italy was able to provide the sole challenge to the United States.

Once again the Italians had full Government support, although this cannot detract from the fact that they designed and built the Macchi M.39 monoplane seaplane in just six months. It did not help the Americans, that they were not having everything their own way. They suffered a series of accidents which lost them some of their best pilots, and problems arose from handling aircraft into which they had fitted a variety of powerplants.

Three M.39s with 800-h.p. Fiat A.S.2 engines were used by the Italian team sent to Hampton Roads. Their pilots were Major Mario de Bernardi, Captain Arturo Ferrarin and Lieutenant Adriano Bacula. The United States team, drawn entirely from the United States Navy, used three Curtiss biplanes: the R3C-2 with a 600-h.p. Curtiss V-1400 engine, flown by Lieutenant Frank C. Schilt, a Curtiss R3C-3, flown by Lieutenant Walter G. Tomlinson and using a 700-h.p. Packard 1A-V-1500 engine, and a 700-h.p. Curtiss V-1550-powered R3C-4 flown by Lieutenant George T. Cuddihy. A heavy landing during the navigability tests caused the loss of Tomlinson's R3C-3, and the only available substitute was a standard Curtiss Hawk float-plane.

The strength of the Hawk was its reliability; it could be counted on to finish the contest, which was more than the temperamental racing machines could offer.

Weather on the day of the 1926 contest was perfect, even though it was late in the year. The first aircraft took off just after the starting time of 2.30 p.m. Both teams suffered aircraft damage: Cuddihy in the Curtiss R3C-4 retired before the end of the final lap with a fuel pump failure, and long before this Ferrarin had been forced to retire during his fourth lap with an oil pipe fracture. Bernardi's winning average speed of 246.50 mph was comfortably ahead of Schilt's 231.36 mph, while Bacula was lagging behind the leaders with 218 mph and Tomlinson's Hawk displayed the performance difference between service and racing machines with a lingering 137 mph!

Italy had regained the Schneider Trophy, and the monoplane had returned to the predominant position in high-speed flight. However, Italian determination to retain the Trophy had to face a British challenge, backed for the first time by the British Government with Royal Air Force personnel flying and maintaining the machines.

The 1927 contest reverted to a bilateral rather than a truly international event. As at Hampton Roads the course consisted of seven laps on a 50-kilometre circuit. Venice was the venue again.

The end of Government support for the Americans had been a not unexpected blow to that country's hopes of regaining the Schneider Trophy in 1927. The last straw was the failure of the private-venture Kirkham-Packard design to produce a worthwhile performance during tests in America. In the face of Italian and British opposition to a month's postponement of the contest, the United States had to withdraw.

The Italians were banking on a development of the M.39, the M.52, with a 1,000-h.p. Fiat A.S.3 engine, to defend the Trophy successfully. Three of these aircraft were ready, to be flown by Major Mario de Bernardi and Captain Arturo Ferrarin (the victors of 1926), and Captain Frederico Guazzetti. Guazzetti's machine's A.S.3 was replaced by an A.S.2, giving Italy one reliable machine, a move prompted by problems with the up-rated A.S.3.

Great Britain's challenge was mounted by Flight Lieutenants S. N. Webster and O. E. Worsley, R.A.F., flying, respectively, propeller-geared and ungeared versions of the Supermarine S.5 seaplane, a braced cantilever development of the ill-fated S.4 with an 875-h.p. Napier Lion engine. A third British aircraft was the biplane Gloster IVB, to be flown by Flight Lieutenant S. M. Kinkead, R.A.F. The Gloster also used a Napier Lion engine, driving the propeller through a reduction gear. A third British type, the Short Crusader monoplane, with an air-cooled radial engine, was taken to Italy for further evaluation, but eliminated itself well before the contest by crashing as a result of having been reassembled with the aileron controls reversed; fortunately, the pilot escaped serious injury.

After a twenty-four hour delay due to bad weather, the contest finally got under way at 2.30 p.m. on Monday, 26 September. The Italian team was fated to lose from the start, as the unreliability of the up-rated engines forced Bernardi to retire during his second lap and Ferrarin during his first. Even Guazzetti, with his 'reliable' aircraft, was forced down after five laps with a broken fuel line. Only Kinkead, who had been doing sufficiently well at one stage to have challenged the leading S.5, had to retire from the British team, with a cracked spinner.

This sad tale of retirements and dashed hopes nevertheless meant a happy ending for Webster, who won the Trophy for Great Britain with an average speed of 281.65 mph. Worsley followed with a speed of 273.07 mph. Webster had lost count of the number of laps flown and discovered on landing that he had flown an extra lap!

Although based on Calshot, at the mouth of Southampton Water, the 1929 contest actually had its start and finish lines 'drawn' at Ryde, on the Isle of Wight. As in previous years, seven 50-kilometre laps were required from the competitors, and taxying tests were added to the mooring-out trial.

Again, this was to be another two-nation event, although by accident rather than design. The Italians, taking a leaf out of Britain's book, attempted to obtain a number of competing designs from which a Schneider challenger or challengers could be selected. But they squandered scant know-how amongst Macchi, Fiat, Savoia-Marchetti and Piaggio and, in the end, could still only race the Macchi design. The French were forced to withdraw because of a number of serious accidents during training for the contest in France, and in the United States poor Alford Williams was still plagued by technical problems. Not that everything went as expected for the British, whose Gloster VI monoplane had to be withdrawn because of problems with the fuel feed system to its Napier engine.

In spite of such widespread trials and tribulations, both the United Kingdom and Italy were able to field complete teams. Britain once again had Government support and R.A.F. personnel for the contest, with Flying Officers H. R. Waghorn and R. L. R. Atcherley, R.A.F., both using 1,900-h.p. Rolls-Royce R-powered Supermarine S.6s, developed from the S.5. One of the latter was entered to be flown by Flight Lieutenant D. D'Arcy Grieg, R.A.F. The Italian team consisted of two Macchi M.67s with 1,800-h.p. Isotta-Fraschini engines, to be flown by Lieutenants Remo Cadringher and Giovanni Monti, and a modified Macchi M.52R with a 1,000-h.p. A.S.3 engine to be flown by Warrant Officer Dal Molin.

On 7 September, the day of the contest, the excellent weather of the preceding few weeks continued, but so did the bad luck of the Italian team. Even with their engines running at reduced power, both the M.67s had to retire during their second laps, their pilots temporarily overcome by smoke and fumes. Another unfortunate was Atcherley, who lost his goggles on take-off and, with his head consequently well down inside the cockpit, took a pylon on the inside and was disqualified, although his speed records for 50 kilometres and 100 kilometres were allowed to stand.

Waghorn won the contest convincingly with an average speed of 328.63 mph, against Dal Molin's 284.4 mph and Grieg's 282.11 mph. Before his disqualification, Atcherley had achieved an average speed of 325.54 mph. The result was a pleasant surprise for Waghorn, who ran out of fuel on his last lap and landed thinking that Italy had won, only to discover that he had repeated Webster's mistake of 1927 and flown an extra lap! However, Italy would have regained the Trophy had Waghorn failed Britain, since the S.5 was not an adequate match for the Macchi M.52R.

The 1931 Schneider Trophy contest was an anti-climax. Plagued by troubles on their new Macchi-Castoldi MC.72, the Italians could not compete. This left only the British team, and on Sunday, 13 September 1931, again at Calshot and over the 1929 course, Flight Lieutenant J. N. Boothman, R.A.F., flew a Supermarine S.6B with a 2,350-h.p. Rolls-Royce R engine to win the Trophy outright for the United Kingdom with an average speed of 340.08 mph. Later the same day, Flight Lieutenant G. H. Stainforth took the S.6B to a world air-speed record of 379 mph.

The real drama for the 1931 Schneider Trophy was political and financial. Despite a promise in 1929 that the Government would support another British attempt on the Trophy, the new Labour Government refused to provide financial assistance. A financial crisis was blamed for this short-sighted attitude, although the British Labour Party has never been notable for understanding the needs and the value of the country's aircraft industry. Promises of funds by the aircraft industry and other businesses fell a long way short of the amount required, and it was not until Lady Houston, the widow of a wealthy shipping magnate, offered £100,000 towards the cost of Britain's entry, that the British team was able to make a start on its preparations for the event. It was another struggle to obtain the services of R.A.F. personnel for a third event, but this too succeeded.

In one respect, the Italians had the final word. On 23 October 1934, Warrant Officer Francesco Agello flew the Macchi-Castoldi MC.72, with its contra-rotating propellers, to an as yet unbeaten record for propeller seaplanes of 440.68 mph. Yet, from the S.6B, Mitchell, its designer, was able to develop the Supermarine Spitfire fighter, which formed the backbone of Britain's fighter defences during the Battle of Britain; thus the United Kingdom gained something lasting from the contests, apart from the Trophy. It is a sobering thought that the Spitfire owes its direct and illustrious ancestry to a generous and patriotic gesture, without which Mitchell would have been denied his own contribution.

Jacques Schneider, born in January 1879, died in 1928. Nevertheless, although he failed to see the final glory of the contest inspired by his Trophy, on more than one occasion during his life the Trophy contest was won by aircraft which had a profound bearing on aeronautical development.

The Trophy itself is on display to this day at the United Service and Royal Aero Club in Pall Mall, London.

Jacques Schneider, who instigated the Schneider Trophy seaplane contests by the presentation of the Trophy in December 1912. He believed that the seaplane, or hydro-aeroplane as it was then called, represented the most potential for speed in the air; although overall this assessment was incorrect, it is nevertheless true that between the two world wars the absolute speed record was frequently held by seaplanes. Although coming from a wealthy family, he died in reduced circumstances in 1928. (Photo: Musée de l'Air, Paris.)

Flying at a time when the Americans were failing to maintain their earlier lead, Charles Weyman used a Nieuport monoplane in the first two Schneider Trophy contests, in 1913 and 1914. He is seen here standing behind the cockpit. He was one of the first to apply some thought to float design. His floats were stepped in accordance with hydro-planing experience. In 1914 Weyman was one of the very few to take the Sopwith Tabloid seriously. (Photo: Musée de l'Air, Paris.)

Considered at first to be something of a joke by the French, the Sopwith Tabloid float-plane won the 1914 Schneider Trophy contest with an average speed of just 86·78 mph, using a 100-h.p. Gnome Monosoupape rotary engine. This is a photograph of the aircraft after winning the race. The tail-float was an afterthought in this conversion from the landplane to counter the effects of some tail-heaviness which was most in evidence during taxying and take-off. (Photo: Hawker Siddeley Aviation.)

The only aircraft to compete in the 1920 Schneider Trophy contest was this Savoia S.12 flying-boat, flown by Lieutenant Luigi Bologna of the Marina Militaire Italiana at a sedate 107.22 mph. One reason for the lack of competition was the stipulation, suggested by the Italians, that all competitors should carry 300 kilograms of ballast, a regulation which was dropped for 1921. (Photo: Siai-Marchetti.)

Italy also won the 1921 Schneider Trophy contest, when Giovanni de Briganti flew this Macchi M.7 (a design which first appeared in 1918) at an average speed of 117.9 mph. Another M.7 ran out of fuel, and a much faster M.19 caught fire after the first lap at a brisk 141 mph. (Photo: Aeronautica Macchi.)

The Savoia S.51 flying-boat came a close second to the Supermarine Sea Lion II in the 1922 Schneider contest at Naples, making an average of 143.5 against 145.7 mph. It would have done even better had it not been damaged during an accident before the contest, and in December 1922 it established a flying-boat record of 174.08 mph. (Photo: Siai-Marchetti.)

In every way, the 1923 Schneider Contest at Cowes was a classic. For the first time in the post-World War I history of the contest there was competition, and in the winning Curtiss CR-3 seaplane of Lieutenant David Rittenhouse, U.S.N., technical development took a vast stride forward with an average speed of 177.38 mph. Rittenhouse is shown here standing on one of the floats of his aircraft with two members of the ground crew holding the aircraft steady – and perhaps impatient to be out of the water! (Photo: Smithsonian Institution, Washington.)

Looking very much the part, Lieutenant James Doolittle, U.S.A.A.C., winner of the 1925 Schneider Trophy contest, is shown here standing on one of the floats of his mount, a Curtiss R3C-2 seaplane which won at an average speed of 232.57 mph. The advance in streamlining compared with the CR-3 can clearly be seen, although this also meant an effective reduction in the view from the cockpit. (Photo: Smithsonian Institution, Washington.)

Outright victory was denied the Americans by Italy's win in the 1926 Schneider Trophy contest, when Major Mario de Bernardi, shown half-out of the cockpit of his Macchi M.39, flew the aircraft over the course at an average speed of 246.50 mph. (Photo: Smithsonian Institution, Washington.)

The Schneider Trophy was regained for the United Kingdom in 1927 by the Supermarine S.5, flown by Flight Lieutenant S. N. Webster, R.A.F., at an average speed of 281.65 mph over the Venice course. The aircraft later established a world speed record. In this photograph Webster is standing at the back with one hand on the aircraft. In the middle of the front row is the aircraft's designer, Reginald Mitchell, who designed the S.6, S.6B and Spitfire developments and is regarded by many as Britain's greatest aircraft designer. (Photo: British Aircraft Corporation.)

Testing! The Supermarine S.6B, which won the Trophy outright for the United Kingdom, during pre-contest checks at Calshot. This was the first aircraft to exceed 400 mph, although the contest was flown at an average of just 340.08 mph in 1931. Today, the aircraft is preserved in the Science Museum, London. (Photo: British Aircraft Corporation.)

Chapter 4

RECORDS BETWEEN THE WARS

The seaplane comes to the fore – the Supermarines.

Sadi Lecointe landing in his much modified Nieuport-Delage 'sesquiplan', during trials for the Coupe Deutsch in September 1922. It was at this event that he took the official record over the 200-mph mark for the first time, and did so by a considerable margin. (Photo: Flight International.)

It was some time before fresh attempts on the air speed record could be made after the end of World War I in 1918. This was not so strange as it may seem. Unlike World War II, World War I had not been a war of technological innovation, but rather one of the full exploitation of existing technology. There were also few aircraft ideally suited to a record breaking attempt, as warplanes of the period did not include any particular type which could be seen to be far ahead of its contemporaries.

The last honours of peace had gone to the French, with the Monocoque Deperdussin, and it was to them also that the first honours of the new peace were to come. This was despite the tremendous advances by the British and Italian aircraft industries during the war and the relative decline of the French.

Little is known about the first post-war record, established on 7 February 1920, by Sadi Lecointe, who flew a Nieuport-Delage at 171.05 mph. But it was the first of many such records during the early 1920s by this daring and popular pilot. Indeed, at this time personalities still mattered in aviation, and a strong air of rivalry and ambition soon became apparent between Lecointe and other pilots of his calibre.

Interest in Lecointe's achievement was negligible, possibly because British attention was taken up by the not irrelevant or unexciting London to Cape Town air race, then in its early stages. At the same time, a major reason for the interest of the British press in the Cape Town air race must have been the existence of newspaper-sponsored entries, although, to be fair, the event had a considerable bearing on commercial aviation development.

Only a little more interest was in evidence when Lecointe's fellow country-man, Jean Casale, succeeded in edging the record up to 176 mph, just three weeks later. This was while flying a Spad-Herbemont with a 300-h.p. engine. To be precise, the official record was 176.15 mph, but during pre-record attempt trials Casale had in fact managed to fly at 178 mph.

While the Cape Town air race was of compelling interest, there is evidence that the workings of the London to Paris and London to Amsterdam commercial air services were of more interest to Britain than attempts to better the air speed record! An argument could be made, of course, that the air services were sometimes more exciting and eventful.

Something of the pre-war atmosphere of the fashionable and exciting Rheims aviation meetings was regained at the first major European meeting of the post-war period, the Buc Trials. The parallels with Rheims were certainly there to be drawn. A succession of speed, altitude and related records were established, and the event was honoured by the presence of the new President de la République, with many of his ministers.

One commentator felt moved to mention the 'very fine flying at Buc', and naturally an 'outstanding item was the speed trial over one kilometre', which was then the official record distance. Lecointe's main rival, the Baron Bernard de Romanet, first came to the fore at the Buc Trials. On 9 October he flew in the speed trials in competition with Lecointe and Jean Casale under the surveillance of the Fédération Aéronautique Internationale. The results were well worthwhile. A new official speed record was established by De Romanet flying a Spad at an average speed of 181.87 mph, while Lecointe, who came second, did well to push his Nieuport along at an average of 179 mph, still higher than Casale's record. Poor Casale, who came third, was unable to match his own record, even though flying the same Spad aircraft.

By this time, Lecointe had acquired the burning ambition to be the first man to establish an official record in excess of 300 kilometres per hour, or about 187.5 mph. Concerned that De Romanet might deprive him of this distinction, Lecointe made a further attempt to set a new record later that day, managing almost 183 mph, or 293.877 kmph. The F.A.I. refused to ratify this as an official record because it failed to meet the official criteria of being at least four kilometres per hour in excess of the then official record, a requirement of that time, designed to guarantee against chronological error.

Such minor setbacks are, of course, encouragement to the true pioneer. The next day, 10 October, Lecointe succeeded in setting a new record of 184.36 mph, or 296.64 kmph, thus putting himself back in the lead, if still short of his coveted 300 kmph.

The pre-war atmosphere of successive speed records had definitely returned, and 1920 had almost acquired the bustle which had characterized 1912 and 1913 when the Monocoque Deperdussin had been in its prime.

Earlier in 1920, Kirsch, another Frenchman, had flown a specially modified Nieuport-Delage biplane in that year's Gordon Bennett Cup Race. The main feature of Kirsch's aircraft was the substitution of a thirteen-metre wing for the twelve metres of that on Lecointe's aircraft, although it retained the 300 h.p. Hispano-Suiza engine which had served Sadi Lecointe so well. On Wednesday, 20 October, Sadi Lecointe was able to borrow the Kirsch machine at Villacoublay for another attempt on the record. He made the first of the mandatory two runs over the measured kilometre straight and took 12.1 seconds, but on the return he managed 11.7 seconds, making an average of just 11.9 seconds, or a new record of 189 mph.

The true significance of this 'unround' number was not lost on sympathetic Anglo-Saxon observers, but for the jubilant Sadi it meant that he had fulfilled his ambition by establishing an official air speed record of 302.5 kmph!

This was faster than the winning speed of 156 mph for the first Pulitzer Trophy Race, which had been held during 1920 in the United States. The winning Pulitzer aircraft had been Lieutenant Curtiss Moselyn's Verville-Packard. But in the future the Pulitzer Trophy was to attract many very fast aircraft and to have an effect on the speed record. The Pulitzer had also seen the Dayton-Wright R.B. high-wing racing monoplane, which had an undercarriage which retracted into the fuselage, although at the then current speeds, fairings for the undercarriage would have been lighter and more effective than a retractable undercarriage.

Public interest had been re-awakened by the Buc Trials, and was now kept alive by what was frequently described as the 'speed duel' between the Baron Bernard de Romanet and Sadi Lecointe. Flying once again at Buc on 4 November, De Romanet flew over the measured kilometre in his Spad at an average time of 11.65 seconds, making an average speed of 192 mph. Before this, in one of his trial flights preceding the record attempt, De Romanet managed a speed of 200 mph, which for the British and the Americans was far more significant than 300 kmph. Naturally, speculation was rife on who would be the first to take the official record over the 200 mph mark.

De Romanet's Spad, with its Hispano-Suiza motor, had also been extensively modified, and some of these modifications were more in keeping with subsequent practice than those on the Lecointe and Kirsch aircraft. Observers were much impressed by the appearance of the aircraft, which had been seen earlier in the Gordon Bennett contest and had since been modified. The pilot sat low down in the fuselage, all but covered in and with little forward vision, but directional stability was improved by the fitting of a larger fin, and the engine had had its compression ratio raised. Details were not released, but there was considerable speculation over the landing speed, although the consensus of opinion was that it must be high because of the heavy wing loading of the Spad.

Many felt that De Romanet had been cheated of a 200 mph record by poor visibility on the day of the attempt. If his problems due to the design of the aircraft had not been bad enough, they had been compounded by a heavy mist, and no doubt he would not have flown at all on that day had it not been for the competitive pressures then current.

The emphasis was now very much on streamlining, including reductions in fuselage front area and recessing the hapless pilot's position as far as possible, if not further! From this time on fuselage diameter was also progressively reduced, and their cramped position in the aircraft allowed few competition pilots the opportunity to wear the parachutes then coming into general use. At the same time the altitude at which record and racing flights were flown was so low that a parachute would have often proved to be worse than useless.

On Sunday, 12 December, Sadi Lecointe flew the Gordon Bennet Nieuport-Delage at Villacoublay; the pilot's position was completely recessed within the fuselage for the first time. Flying at a steady, but incredible three metres above the ground, the modified aircraft was capable only of raising the record from 192 mph to 194 mph, while the 200 mph barrier was broken in one direction only. This was the last flight of 1920. No official records were set during 1921, although Sadi Lecointe managed an unofficial 210 mph by the end of the year.

It was Sadi Lecointe's privilege in the end to establish the first official speed record of more than 200 mph, although he was nearly cheated of the distinction. An Italian, Brack-Papa, flew a Fiat biplane, powered by a 700-h.p. Fiat A.V.12 cylinder engine, at 209 mph, but the Italian Aero Club, for some reason best known to itself, refused to recognize this as a record and the F.A.I. did not even have the chance to consider the matter. The Italians were later to insist on impartial observers being present before recognizing any record attempt themselves, and this self-inflicted requirement may not have been without relevance to the Brack-Papa case.

Sadi Lecointe's opportunity came on Thursday, 21 September 1922, at the time of the trials for the Coupe Deutsch, a trophy awarded by the Deutsch de la Meuthe family. His Nieuport had been reduced in size, to the amazement of those who saw it, and fitted with a 320-h.p. Hispano-Suiza engine. By this time, he had to make four runs over the single straight kilometre to establish a speed record. On its first flight, Lecointe's 'sesquiplan' (as he now called it) covered the kilometre in 10.4 seconds, taking 10.6 seconds on the second flight, 10.4 on the third and 10.8 on the fourth, producing an average speed of 211.6 mph.

This record took Lecointe over the 200 mph mark by a considerable margin, and, having taken place under the watchful eyes of officials from the Aéro Club de France, his record attempt, at Étampes, was spared the frustrating lack of recognition which had hindered Brack-Papa.

A more serious competition than that posed by his fellow countrymen or by the Italians was about to threaten Lecointe. The first real indication of this was the victory of a Curtiss Navy Racer biplane, flown by Bert Acosta, in the 1921 Pulitzer Trophy contest at Omaha.

On 13 October an American army officer, Brigadier General William A. Mitchell, flying a Curtiss HS D-12 biplane at Detroit, succeeded in raising the official speed record to 222.98 mph. Mitchell, a controversial and patriotic figure in the United States because of his struggle for an independent military air arm, later tried to raise the record further, but this was never recognized. At about this time too, a record of 248.5 mph was claimed for Lieutenant Russell L. Maughan, winner of the 1922 Pulitzer Trophy in a Curtiss biplane. This record was never confirmed.

Sadi Lecointe's last speed record came on the morning of Thursday, 15 February 1923, in his Nieuport 'sesquiplan' at Istres. Equipped with a 400-h.p. Hispano-Suiza engine the plane made the required four flights over the straight kilometre at an average speed of 233 mph, although on the fastest of these his time for the kilometre was just 9.2 seconds, a speed of more than 244 mph. The aircraft's landing speed of 112 mph astonished most commentators at a time when this was more than the cruising speed of many aircraft. Such high speeds from the technology of the day were having many results, not the least of which was the attempt to minimize drag and engine overheating problems by placing large radiators in the wings of the Curtiss and Nieuport aircraft.

Now it was the turn of the Americans to establish a number of speed records, spurred on in this, as in the Pulitzer and Schneider contests, by rivalry between the U.S. Army and the U.S. Navy.

The first of the new American records was the last to be flown over a single-kilometre straight, after which the three-kilometre rule came into force. At Dayton, Ohio, Lieutenant Lester Maitland raised the official record from 233 mph to 236 mph on 29 March 1923, while flying a Curtiss R-6 biplane. He also managed an unofficial 243 mph.

Poor Sadi sought refuge in setting a new altitude record.

Two days after that year's Pulitzer Trophy Race of 2 November, Lieutenant A. Brow established a new record with an average speed of 257.5 mph in his Curtiss HS D-12 at Mitchell Field, Long Island, New York.

The winner of the 1923 Pulitzer race, the then Ensign Alford J. Williams, U.S.N., who was later to be so unlucky in his efforts to regain the Schneider Seaplane Trophy for the United States, immediately attempted to improve upon Brow's record. Williams's Curtiss R-2C-1 (his victorious Pulitzer machine) was a landplane version of the aircraft which had carried Doolittle to victory in that year's Schneider Trophy contest. However, in spite of this distinguished record, Williams, in the R-2C-1, was unable to raise the speed record beyond the four-kilometre rule to set a new official record.

Anxious to defend his record, which now stood only upon a technicality, Brow at once returned to the fray; in the final official record attempt of that eventful day, he produced a new record of 266.60 mph. An excellent day's work and brilliant flying by any standard, but by the close of 4 November, there were also unofficial records of 270.50 for Williams and then 274.20 mph for Brow.

Almost as if he was trying to detract from the importance of speed, Sadi Lecointe went on to establish yet another altitude record later that month.

The British aircraft industry was conspicuous by its absence from these proceedings, although it had won a Schneider contest since the war. Distance and trailblazing seemed to be the forte of the Briton, even though a highly futuristic racing monoplane with low-wing and retractable undercarriage had been built and tested by Bristol in 1922.

What would otherwise have been a lull of several years without new records was interrupted towards the end of 1924 by the first official Italian speed record. An Italian Air Force Officer, Adjt. A. Bonnet, flew a Barnard-Ferbois biplane at Istres Aerodrome at 278.5 mph on 11 December. This aircraft used a non-Italian powerplant, a 450-h.p. Hispano-Suiza water-cooled unit driving a fixed-pitch metal propeller — still something of an innovation at this time when wooden propellers were the usual order of things.

This was one of the first peacetime records to endure for a reasonable period, but records are only made to be broken. Nevertheless, this one lasted for almost three years, until Flying Officer Sidney Webster, A.F.C., R.A.F., brought the first British speed record back from the 1927 Schneider Trophy contest at Venice. Webster's record of 284.21 mph on 26 September was unofficial, however, and in fact Bonnet's record thus had, as it were, a stay of execution.

Britain had two potential record-breaking aircraft in the 1927 Schneider team. Webster's Supermarine S.5 monoplane seaplane with a geared propeller, which won the contest at an average speed of 281.65 mph, and Flying Officer Kinkead's Gloster IVB biplane, also with a geared propeller, which was challenging the S.5 when it had to withdraw from the contest with a cracked spinner. Another S.5 had an ungeared propeller, and although this aircraft managed to stay the course, it was at an average speed of 273.07 mph — too slow for any attempt on the speed record.

Nevertheless, the significant feature of the S.5's performance could not be denied. For the first time a seaplane had shown itself capable of exceeding landplane speeds. The dream of Jacques Schneider was coming true, and would remain so for some years, although the great man's death was not far away.

Although they had lost the Trophy, the Italians were determined to cling on to the official speed record for as long as possible. Their famous Schneider Trophy pilot, Major Mario de Bernardi, remained at Venice after the contest to fly his Macchi M.52 monoplane seaplane to a new official record of 296.82 mph on 5 November 1927. Determined that their fresh record should be beyond all doubt, the Italians went to great lengths to ensure the presence of impartial observers in the form of the British, American and French air attachés, while the attempt was made under the auspices of the Italian Aero Club.

This was a convincing display by the Italians and a reasonable consolation prize for being deprived of the valued Trophy. Further displays of Italian strength in aeronautics were also planned, the first of which was a further record attempt by de Bernardi at the Lido, Venice, on 27 October, when he reached 298 mph. Italy did not accept this herself, and the F.A.I. did not even have the opportunity to pronounce on the matter. The Italian authorities had by now come to be insistent on the presence of impartial observers. Why the Italians took this attitude is difficult to understand, as it is unlikely that the word of the Italian Aero Club would have been disputed.

A further unofficial record came from the de Bernardi and M.52 combination at the Lido later that November. The new unofficial record carried the Italian monoplane to well over the 300 mph mark at 313.6 mph, although there was less fuss about this than over the 300 kmph or 200 mph marks.

There was much for the Europeans to be concerned about as reports filtered through to Europe about Lieutenant Alford J. Williams reaching a speed of 322.6 mph during trials on 6 November with his Packard-powered Curtiss float-plane, the aircraft which he would have taken to Venice for the Schneider Trophy contest but for delays in development. The concern of the Italians was nothing compared with that of the British, who were trying desperately to kindle official enthusiasm for the Schneider contest in order to be able to compete on a par with the Europeans and Americans.

There was no Schneider contest in 1928 as a result of the agreement that, after 1927, it should be held every other year. Determined to show that they were maintaining a high standard of preparation, and eager to keep the spirit of competition alive, the Italians sought a new record for the Macchi M.52, with an up-rated Fiat AS.3 engine in 1928.

They did not keep their rivals waiting unduly. On 30 March, at Venice, Major de Bernardi established a new official record of 318.5 mph over the three-kilometre straight. As always, the record was only the average of four attempts. An unofficial record of 348.5 mph was established from the fastest time over the straight.

Not unnaturally in this prevailing climate, the British team for the 1929 Schneider Trophy contest were thinking in terms of a speed record as well as another Trophy victory. During the actual contest, in the Solent on 7 September, an unofficial 370 mph was reached by Flying Officers R. D. Waghorn and R. L. R. Atcherley, R.A.F., in their 1,900-h.p. Rolls-Royce 'R'-engined Supermarine S.6 monoplanes. Waghorn in fact won the contest, while the unfortunate Atcherley was disqualified for taking a pylon on the wrong side. He had had to keep his head well down inside the cockpit after losing his goggles during take-off.

The period of excellent weather which covered the contest and its preliminaries continued after the event, giving the British team sufficient time for an attack on the official air speed record. A measured three-kilometre straight in Southampton Water, not far from the team's base at Calshot, was used, over which the captain of the team, Squadron Leader A. H. Orlebar, R.A.F., had already flown at 355.8 mph in Waghorn's S.6.

Orlebar took-off from a smooth sea at 4 p.m. on 12 September and flew over the straight six times. His four consecutive best times were on the second, third, fourth and fifth flights. He reached speeds of 354.6 mph, 358.7 mph, 352.5 mph and 365.1 mph, respectively – an average of 357.7 mph, the new official record. Photographic timing was used for the first time on this record attempt.

The interval between the 1929 and 1931 Schneider Trophy contests saw a struggle for official support to be continued for the British team. This type of support was taken for granted by other European participants, and it had a bearing on the ability to mount an attempt on the speed record. In the end, the British team went forward with the support of private patronage.

Fortunately, too, for the British, their rivals were unable to develop comparable aircraft in the time allowed.

The 1931 Schneider Trophy contest was won by the British team for the third time running, with a 2,354-h.p. Rolls-Royce 'R' engine in a Supermarine S.6B flown by Flight Lieutenant J. N. Boothman, R.A.F. The plane averaged 340.08 mph over the course on Sunday, 13 September. Later that day Flight Lieutenant G. H. Stainforth flew over a three-kilometre straight between Lee-on-Solent and Hillhead, establishing a new speed record of 379 mph, in the reserve S.6B.

A further record remained to be set by Stainforth on Tuesday, 29 September. After waiting for early morning mist to clear, he once again flew over the three-kilometre straight at an average speed of 408.8 mph in the S.6B. Partly thanks to ideal conditions with just a slight breeze to take the glassiness off the sea, an unofficial record of 415.2 mph was set during the first flight; the third flight was at 409.5 mph, and the second and fourth at 405.1 mph and 405.4 mph, respectively.

This gave to the United Kingdom the distinction of the first flight at more than 400 mph. It also placed the record and racing machines some ten years ahead of the fighter aircraft in terms of speed.

Once again the Italians were determined not to suffer defeat without some mitigating display of technological merit – although it had been some defeat, as Britain won the Schneider Trophy outright after Italy was unable even to field a rival team! It was not until 1933 that the next attempt on the speed record could be made.

The Italian Schneider Trophy Team had been kept intact as the High Speed Flight, under the continuing command of Colonel Bernascari. It was one of the more junior members of the flight, Warrant Officer Francesco Agello, who at

Lake Garda on 10 April 1933, flew a Macchi-Castoldi MC-72 seaplane to a new record of 423.08 mph. A year later Agello was back at the same place waiting for the weather to clear for another attempt at the record, but the only record of that spring was an unofficial 434.96 mph by the faithful Bernascari.

Italy's real triumph had to wait until the autumn of 1934. Tuesday, 23 October, saw Warrant Officer Agello on Lake Garda once more with his 3,000-h.p. Fiat 24 cylinder-powered Macchi-Castoldi MC-72, determined to raise the record still further.

In contrast to the spring, the weather was ideal, with just enough movement on the surface of the water to aid take-off, while the air was free from bumps. Just before 3 p.m. Agello took off for his four flights over the measured three-kilometre straight, which ran north to south by the lakeside. The result has still to be bettered, for he succeeded in establishing an all-time speed record for propeller seaplanes of 440.68 mph, as well as a world air-speed record. A good indication of the conditions prevailing at the time can be taken from the fact that his slowest flight was at 438.6 mph and the fastest at just 442.081 mph. A small difference indeed!

Almost five years were to elapse before the next official air-speed records moved Agello and the MC-72 down the ladder of history. In the intervening period the political situation in Europe changed considerably, to the extent that information on the last records of the inter-war period proved to be relatively scant. This was no longer speed for sport, or even for technological progress, but rather speed for propaganda, with many interesting details suppressed in the interests of secrecy.

The first of these records was established in the accepted manner by Flugkapitan Hans Dieterle of the Luftwaffe on 30 March 1939 while flying a specially-modified Heinkel He.112V at Oranienburg, near Berlin. For the first time in almost fifteen years the speed record returned to a landplane; it was a significant advance over earlier records at 463.92 mph. Relatively little is known about the attempt, although the aircraft did use a 1,175-h.p. Mercedes-Benz water-cooled engine and had a distinguished record: it had been flown by the ace, General Udet, to a speed record over 100 kilometres (62.13 miles) the previous year.

After such a long interlude, it is strange that the next speed record should follow within a month. Stranger still that it should be another Luftwaffe officer flying a different aircraft type, Flugkapitan Fritz Wendel in a specially modified Messerschmitt Bf.109R fighter from the Messerschmitt airfield at Augsburg. In contrast to the 950 h.p. of the then standard Bf.109 engine, the Bf.109R used the same 1,175-h.p. Mercedes-Benz engine as the Heinkel, but the Messer-schmitt made better use of the power, setting the last official speed record of the inter-war period with 469.22 mph on 26 April. The record was only slightly better than that of the Heinkel but almost 100 mph better than the maximum speed of the production Bf.109 at the time.

Three days later Wendel took the Bf.109R to an unofficial record of 486 mph.

Although faster piston-engined aircraft were developed during the war, and production aircraft were able to match Wendel's records, no further attempts were made to set an official speed record using piston-engined, or indeed turboprop, aircraft for another thirty years. By the time Darryl Greenamyer flew his Grumman F8F-2 Bearcat at 776.499 kmph (485 mph) on 16 August 1969, it could no longer qualify as an absolute speed record; although it was certified as a new piston-engine record.

In nineteen years the speed record had risen by some 300 mph, more than twice the increase of the first seventeen years of heavier-than-air powered flight. In the process the shape of the aeroplane had changed, thanks to extensive streamlining, the general move to the monoplane configuration, changes in construction, and the replacement of the wooden propeller by the metal propeller and then the variable-pitch propeller, which offered optimum settings for take-off and for cruising flight.

Against these gains, for gains they must be when the object of the exercise is speed, must be set the not entirely welcome fact that speed records were no longer the preserve of the individual. Indeed, a large fortune could be thrown away in an attempt on the record. Nor was speed within the resources of most manufacturers. Government support was, by the late 1930s, almost imperative for any attempt on the speed record; indeed, the Americans and the British had discovered that it was advisable to have government support as far back as the 1920s, if only to counter French and Italian tactics.

Racing aeroplanes could no longer be really fast, if the individual was to be able to afford the sport. The days when contests such as the Schneider Trophy, the Pulitzer Trophy and the Gordon Bennett Cup, or even lesser contests such as the Thompson Trophy or the Coupe Deutsch, could contribute towards raising aircraft speeds were gone.

From now on, there are few specifically record-breaking aircraft types. Most of the aircraft are special developments of standard fighter and interceptor designs, and the final record of the inter-war period was a clear pointer to this trend. Although taking much of the colour and glamour out of the record attempts, this trend should not be condemned; not only would speeds continue to rise anyway, but, after all, the main rôle of the fighter is defence.

Winner of the 1923 Pulitzer Trophy Race, the then Ensign, later Lieutenant Alford J. Williams, U.S.N., by his Curtiss R-2C-1. His somewhat grim expression is easily accounted for. His attempt on the speed record had failed by falling foul of the four-kilometre rule. (Photo: U.S.A.F.)

Line up for the 1923 Pulitzer Trophy Race on 2 November. Nearest the camera is Alford Williams's aircraft, No. 10, in which he won the contest but failed to take the speed record; next to that is Lieutenant H. J. Brow's No. 9, also a Curtiss R-2C-1, which came second in the race and allowed him to win the record. The other two aircraft are U.S. Army Wrights. (Photo: U.S.A.F.)

Still the fastest! The Macchi-Castoldi MC-72 seaplane, which established a still unbeaten speed record for propeller-driven seaplanes of 440.68 mph in October 1934. The record was also an absolute speed record at the time. In this photograph, the contra-rotating propellers, then a novelty, can be seen clearly. (Photo: Aeronautica Macchi.)

The last speed record of the inter-war period, and for thirty years the fastest official record for a piston-engined aircraft, was set by this Messerschmitt Bf.109R in 1939. The differences between the record machine and the operational aircraft can be seen to be considerable, with wings which are even more clipped, tailplane, engine cowling and cockpit all different, as was the specially-prepared engine. (Photo: Messerschmitt Archive.)

Well done! Professor Willy Messerschmitt congratulates Flugkapitan Fritz Wendel after his record flight on 26 April 1939. (Photo: Messerschmitt Archive.)

Sir Henry Royce, one of the founders of the famous aeroengine and motor manufacturer, Rolls-Royce. Royce was the engineer, and the Hon. Charles Rolls the salesman. Royce masterminded the design of the 'R' engine used in the Supermarine S.6B which ultimately led to the famous Merlin used in many World War II aircraft, including the Spitfire, Mosquito, Lancaster and Mustang. (Photo: Science Museum, London.)

Chapter 5

THE AIRLINER BETWEEN THE WARS

Converted bombers – development of the first airliners – the monoplane takes over – design for speed – the modern airliner.

The peak of inter-war British civil aircraft development was represented by the de Havilland Albatross, a graceful airliner which entered service with Imperial Airways in late 1938. The prototype is seen here. Other British manufacturers, including Shorts and Fairey, were less lucky and did not even get their aircraft into the air before the outbreak of war brought civil development to a halt. (Photo: Flight International.)

Airline speeds represent the true meaning of speed in the air for the average person, with speed records in the absolute sense appearing as a novelty. In the past, racing and record-breaking aircraft have done much to advance technology, but for the air-traveller standards of safety and reliability must be placed before any other consideration. Economy of operation, which places the emphasis on cruising speed rather than on maximum speeds is also important, and range can be a decisive factor. Refuelling stops can cancel out the speed advantage of one aircraft over another. Nevertheless, progress in airliner speeds is an indication of aviation advancement.

The airliner did not appear until after World War I, even though to describe some of the early passenger aircraft as airliners is to flatter them. A British company, Sopwith, did build a three-seat aircraft before the war, and Russia's Igor Sikorsky built his 'Bolshoi' in 1913, following this with his 'Ilya Mourametz', appropriately meaning 'the giant'. Although the 'Ilya Mourametz' was designed to be an airliner with a number of advanced features, its operational rôle was that of a successful bomber. Most of the early airliners were to undergo the opposite transformation.

Air transport was not a new idea, dating at least from the time of Sir George Cayley (1773–1857). In pre-war Germany, large Zeppelin dirigibles had carried fare-paying passengers on domestic air services.

The paradox of the early years of air transport was that the initial trend was for aircraft speeds to decline, as converted bombers were replaced by aircraft designed to be airliners. These offered better passenger accommodation and superior operating economics.

Although a converted Farman Goliath bomber was used on a tentative Paris–London air service early in 1919, the first regularly scheduled air service in the world was inaugurated with a flight from London to Paris on 25 August 1919, using a converted Airco D.H.4a bomber belonging to the operator, Aircraft Transport and Travel. A single-engined biplane with accommodation for one pilot and two passengers, the D.H.4a had a cruising speed of 100 mph and a range of up to 375 miles. Before long it was joined by another converted single-engined biplane bomber, the D.H.9, which had similar accommodation, a cruising speed of 95 mph, and a range of 380 miles.

The conversion from the bomber rôle included the provision of a cabin for the passengers, but even with this luxury the wise air traveller would wear warm clothes during chilly weather. Cabin service was still some time away, but the airline could usually be relied upon to provide a warm blanket and a hot-water bottle! Fares were just a little lower than today, at around £21 single London to Paris, although they soon fell to 6gns., and in 1927 to £9 return.

If the flight went according to plan, the London to Paris trip would take just

E

over two hours, but efficiency was assessed not on the proportion of flights arriving on time but on whether or not the flight arrived at all. On one occasion, an aircraft took two days and made 33 landings on the journey. Landings were often required because of engine failure or the pilot's getting lost in poor visibility. Instruments and navigational aids were rudimentary.

Overall, the impression of these early years must be that speed was not the only direction in which an improvement was desirable.

Development away from aircraft more suited for joy rides and pleasure flights was not too long in coming, although the bomber, or, to be more exact, the heavy bomber, retained its connection with civil aviation. The Handley Page 0/400 was the first such British aircraft to undergo this conversion: its two 360-h.p. Rolls-Royce Eagle engines enabled it to carry up to fourteen passengers and cruise at 85 mph. The 0/400 was soon joined by the Vickers Vimy, another bomber in civilian guise which had made the first non-stop trans-Atlantic flight in 1919, taking 16 hours, 28 minutes to landfall on the west coast of Ireland. Civil versions of the Vimy retained the engines and aerodynamic surfaces of the bomber but had a large and bulbous fuselage in which the fourteen passengers could enjoy the luxury of basket chairs while the pilot and co-pilot were left out in the cold! The Vimy bomber's speed of some 120 mph dropped to a little under 100 mph as an airliner.

The mighty American civil aviation industry, which today makes that country something of a mecca for airliners and airlines, could at this time do little more than licence-build D.H.4a and D.H.9 biplanes, although the American Liberty engine was often substituted for the original.

A more forward-looking design of this time was the German Professor Hugo Junkers's F.13, a low-wing cantilever monoplane with a corrugated all-metal form of construction. This type of construction was favoured by Junkers for many years and also influenced a number of other manufacturers, notably Ford in the United States. First introduced in 1919, the F.13 was built in both landplane and float-plane versions, with a crew of two and four passengers accommodated in the small cabin. However, the 185-h.p. engine could only give a speed of 90 mph while cruising, which was no better than its biplane contemporaries of a more traditional construction.

Cruising speeds of around 90 mph must be regarded as an early peak for the airliner designers, and from which they had some difficulty in breaking away. Indeed, for many designers it was something to aspire to. Anthony Fokker's chief designer, the German Reinhold Platz, designed the first of the highly successful Fokker high-wing monoplanes in 1920, with cantilever wooden wing and a pilot and four passengers in the cabin; but even with the same 185-h.p. engine as the F.13 a cruising speed of 75 mph was the limit.

German and Dutch designers were pressing forward, although the Germans were largely forced to work in the exile which resulted from the Armistice ban on aircraft production in Germany. The French had surrendered their commanding position in aviation development, and the United States was just beginning that country's rapid rise to the forefront of aeronautics. Too many British designers were in a rut, often being content merely to substitute metal construction for wood and making no accompanying advance in the design of the airframe. To be fair to the British, however, it must be remembered that airline and military purse strings were tightly drawn. There was also a preoccupation with the slower flying-boat for the Empire routes.

The new British aircraft, the biplane Handley Page W8b and its successor, the W10, the Supermarine Sea Eagle amphibian, the trimotor Armstrong Whitworth Argosy and de Havilland D.H.66 Hercules, the Short Calcutta flying-boat of 1928, and even the Handley Page H.P.42/45 Heracles four-engined biplane of 1931, all failed to make any real increase in cruising speeds. On the credit side, many of these aircraft were safe and able to offer an improving standard of comfort for up to 38 passengers (in the case of the H.P.42/45, which also had a 300-mile range).

It was during the mid-1920s that the American aircraft industry started to establish the first of the airliner fashions and to attract the attention not only of America's own airlines but of their more sophisticated European counterparts. Much of the impetus for this development came with the award of air-mail contracts to American airlines. The small, somewhat fighter-like aircraft produced for this were named 'speed planes', appropriately enough.

One of the first of this new generation of aircraft was the Lockheed Vega, a single-engined high wing monoplane which first appeared in 1927 and could accommodate up to six passengers. Early versions used a 220-h.p. Wright engine giving a cruising speed of 100 mph and a range of up to 500 miles, but later models used a 425-h.p. radial engine capable of giving 135 mph cruising speed and a range of around 900 miles. Stressed skin construction reduced fuselage diameter and weight, reducing drag and marking a significant step forward in the development of larger aircraft. Similar in concept were the Lockheed Air Express of 1928, with a cruising speed of 150 mph, and the low-wing Lockheed Orion I of 1930, with a cruising speed of 200 mph and a maximum speed of 225 mph.

Competition came from Boeing biplanes, which placed a greater emphasis on the mail than on the passengers, and later from Northrop, which introduced the Alpha in 1931 and followed this with the Delta and Gamma.

At the same time, a slower and more dependable class of airliner was being evolved in the United States and in Europe. By the mid-1920s Fokker had a factory in the United States, but it is unlikely that a competitor to Fokker would have developed as it did had not Henry Ford bought William Stout's aircraft factory and designs and moved the factory to Detroit. Stout had developed a six-passenger high-wing monoplane using a single 420-h.p. Liberty engine, which gave a maximum cruising speed of 110 mph when it was first introduced in 1924. On receiving backing from Ford, Stout was able to develop this into a family of larger aircraft with one, two or three engines and Junkers-style corrugated metal construction, earning the famous Ford Trimotor the name of 'Tin Goose'.

Boeing also built a trimotor airliner during the late 1920s. This was the 80A, with a cruising speed of 125 mph and accommodation for fourteen passengers. This aircraft also represents one of the earliest attempts to provide pilots on the larger aircraft with an enclosed flight deck. The innovation was not always welcomed!

In Germany, the Junkers G.31 sixteen-passenger transport obtained a cruising speed of 100 mph from its three 450-h.p. radial engines. The Dutch Fokker F.VII of 1924 with a single engine was developed into the trimotor F.VII-3M of 1926, using a variety of powerplants according to availability and customer requirements, but usually only giving a cruising speed of around 110 mph. It was a Fokker F.VII-3M which made the first flights over the North Pole, in 1926, and the South Pole, in 1929, in each case with Lieutenant-Commander Richard Evelyn Byrd, U.S.N. at the controls.

These were the plodding aircraft of the period, but they pioneered many air routes, established endurance records and set point-to-point speed records. They were not fast: even the Fokker F.IX of 1931 could only make a cruising speed of 115 mph. However, these were the aircraft which started the trend towards mass air travel, even though at first this was for the fortunate few, mainly businessmen and diplomats.

Passenger comfort left much to be desired, even with in-flight cabin service. One duty for a steward on a Ford Trimotor, or 'Tin Goose', was to offer passengers cotton wool with which to plug their ears and ease the misery of the high interior noise level. This thoughtful service left them with the discomfort of the seats, which were leather-covered and either basket or metal-backed, and, according to one commentator, compared with garden furniture for comfort! To add a final insult to injury, on landing on a muddy or puddled runway, mud and spray thrown up by the wheels entered the aircraft in liberal amounts through the ventilation vents!

Fortunately, the early 1930s witnessed a significant improvement in all aspects of airliner development and operation and the birth of the modern airliner.

For Britain's Imperial Airways the first significant increase in airliner speeds did not come until 1932 with the introduction to service of the Armstrong Whitworth Atalanta. A small four-engined high-wing monoplane, the Atalanta could cruise at up to 130 mph on its four 340-h.p. engines, although its range was still no more than 400 miles and accommodation limited to nine passengers.

The biplane era was far from over. In the United States, Curtiss was able to build and sell the Condor, a biplane airliner which also introduced the first sleeper flights. In the United Kingdom, Imperial's rival airline, British Airways, introduced the de Havilland D.H.89 Dragon Rapide, a twin-engined biplane intended primarily for feeder services but which was capable of carrying eight passengers for up to 550 miles at a cruising speed of 130 mph. This was in 1934. That same year, the largest of the Dragon series, the four-engined de Havilland D.H.86A Dragon Express was able to carry ten passengers for up to 450 miles at a cruising speed of 145 mph. A landplane operator, British Airways, was able to acquire a faster fleet and a more go-ahead image than Imperial Airways.

The lead had definitely passed to the United States by this time, with most of the emphasis being placed on the landplane rather than the flying-boat, in spite of notable designs in the latter field by Martin, Sikorsky and even Boeing. In order to succeed against railway competition, the American airlines had to keep flying passengers through their routes until the destination was reached, thus forcing the pace in developing faster aircraft, longer range and better navigational aids. European airlines, often with a concentration of effort on Empire routes, tended to make overnight stops. These differences were beginning to tell. Even so, the one-week rail and air journey from the United Kingdom to India in 1929 was a gain over the three-week sea voyage. In 1934, it became possible to travel to Australia in a joint Imperial–Qantas service, taking $12\frac{1}{2}$ days, against six weeks by sea, and costing £195 single.

The birth of the modern airliner can be credited to the Boeing Company, which in February 1933 first flew its new 247 twin-engined, low-wing monoplane airliner. Characteristics of the aircraft were the use of all-metal monocoque construction, cowled and air-cooled supercharged engines mounted in the wings, and a retractable undercarriage to reduce drag in flight. While the capacity of ten passengers was certainly well below that of many contemporary designs, comfort and speed were much improved: early versions had a cruising speed of 165 mph from the two 550-h.p. Pratt and Whitney Wasp radial engines. A later aircraft, the 247D, made 180 mph and introduced variable pitch propellers and wing de-icing equipment.

The Boeing 247 was a civil development of the company's B-9 bomber of 1931. This was not the last time that Boeing would base a successful civil design on military experience.

America's other aircraft manufacturers, and the airlines, were determined not to let Boeing have a monopoly of advanced airliner development. This attitude has remained to the present day, with beneficial results.

By the early 1930s, Boeing was relatively well established, which was more than could be said for the still comparatively new Douglas aircraft. Yet the Douglas DC-1 (Douglas Commercial-1), which was only built in prototype form, was to have a greater impact on air transport than the Boeing 247. Built to meet a specification set by Transcontinental and Western Air (the Trans World Airlines of today), the DC-1 first flew in July 1933. The immediate potential of the aircraft was not lost on T.W.A., which opted for a developed version, the DC-2, with accommodation for fourteen passengers, a range of 350 miles, and a cruising speed of 185 mph from its two 710-h.p. Wright Cyclone radial engines.

Swissair had already bought Lockheed Orions, but both this airline and K.L.M. Royal Dutch Airlines ordered the DC-2. What followed was almost a repeat of the Wright brothers' achievement in France in 1908. European aircraft manufacturers were cynical about K.L.M.'s entry of a DC-2 for the 1934 London to Melbourne Air Race, sponsored by the *Daily Mail* and held in October. Yet, operating the aircraft with a full load of passengers and mail, the DC-2 covered the distance in just three days and eighteen hours, a time second only to that of the de Havilland D.H.88 Comet racing monoplane, which carried no payload! A Boeing 247D came third.

European backwardness was once again convincingly demonstrated, but, with history repeating itself, the Europeans were again spurred into action. They produced a number of elegant and fast modern airliners, starting with the small Junkers Ju.86 twin-engined monoplane of 1936.

Fastest of all the twin-engined all-metal American monoplanes was Lockheed's Electra I. Although not so successful as the DC-2 and not of such epoch-making character as the 247 and 247D, the Electra I was a success for the manufacturer and for its operators, mainly in the United States and Europe. It included all the improvements introduced by Boeing and Douglas, and, like the DC-2, it was the first of a family of aircraft. Amongst the operators were Swissair and British Airways.

Early versions of the Electra, known as the Lockheed 10 Electra, had a cruising speed of 160 mph from two 450-h.p. Pratt and Whitney Wasp radials which also restricted it to just eight passengers. But by 1937 the Lockheed 14 Electra had raised the cruising speed to 225 mph and passenger accommodation to twelve, while range was a useful 1,000 miles.

By contrast, the French Dewoitine D.332 trimotor airliner of 1933 had a fixed undercarriage, which it retained until the D.338 of 1937. This large aircraft's cruising speed was still a useful 160 mph.

Few of the German Heinkel He.111 airliners actually entered airline service, as the aircraft was a thinly disguised bomber type. However, the Junkers Ju.52/3M, which marked the ultimate in the Junkers system of corrugated metal construction, made a major contribution to civil and military air transport for many years, including the period after World War II. Three 575-h.p. B.M.W. radial engines powered the Ju.52/3M when the first versions appeared in 1932, enabling it to carry about fourteen passengers at cruising speeds of up to 145 mph.

By this time Fokker had developed its trimotor series to the extent of obtaining a 155 mph cruising speed out of the fourteen passenger F.XVIII.

If these efforts were intended to match aircraft of the calibre of the DC-2, the Boeing 247 and the Electra, they were to be completely eclipsed by further progress in the United States. Douglas, having found a successful concept in the DC-2, improved upon this with the larger and slightly faster DC-3, which first flew in December 1935. The Douglas DC-3 remained in production from shortly afterwards until the end of World War II. Some 13,000 DC-3s and the related military C-47 Dakotas were built, a record which is still unbeaten. The DC-3 production record excludes aircraft built in the Soviet Union as the Lisunov Li-2.

Two 1,000-h.p. Wright Cyclone piston engines gave the DC-3 a maximum speed of 225 mph and a cruising speed of 170 mph, and the normal range was up to 500 miles. Some improvement on performance was obtained during the production life of the aircraft, but the increase in passenger-carrying capacity from the original twenty-one to thirty-two or more was probably not one of the improvements and probably goes some way towards explaining the poor opinion some air travellers have regarding the DC-3 today. Perhaps the most luxurious version was the fourteen-berth sleeper aircraft, known as the D.S.T. or Douglas Sleeper Transport, of the middle and late 1930s.

No aircraft did so much as the DC-3 to introduce mass air travel. Many of the first cheap post-World War II charter flights used this aircraft. But the Electra cut the London–Paris flight to 1 hour 20 minutes, giving British Airways a go-ahead image. A weekend in Paris flying by Electra cost just six guineas return.

It was no mean achievement for the Short Empire flying-boat of 1936 to attain a cruising speed of 165 mph, which was not too far short of contemporary landplane speeds. True, the large Empire flying-boat normally only carried about twenty-four passengers, but they travelled in superlative comfort and the aircraft also carried a considerable volume of mail and some cargo.

Having developed the twin-engined airliner to what could, for the time being, be considered its peak, the aircraft manufacturers turned their attention to larger four-engined aircraft with a longer range. The European manufacturers were caught less unaware on this occasion; if World War II hadn't intervened they could probably have given the American manufacturers rather more vigorous competition than actually materialized.

In Germany, Junkers developed the Ju.90 and Focke-Wulf the Fw.200 Condor. Great Britain's contribution included the de Havilland Albatross and the Armstrong Whitworth Ensign, although Short Brothers and Fairey both had modern four-engined landplanes under development at the outbreak of World War II. Fokker in the Netherlands had the F.22. The American aircraft manufacturers had the Boeing 307 Stratoliner, the Douglas DC-4 and the Boeing 317 flying-boat.

The first of this new generation of aircraft to become airborne was the Armstrong Whitworth Ensign, which first flew on 24 January 1938. A large, high-wing monoplane with four 850-h.p. engines, the Ensign was at that time the largest all-metal monocoque airliner, able to carry up to forty passengers over distances of 800 miles and cruise at 170 mph (the maximum speed was in excess of 200 mph). The fourteen Ensigns flew with Imperial Airways before passing to that airline's successor, B.O.A.C. (British Overseas Airways Corporation), and undertaking wartime V.I.P. and official communications flights.

The other British airliner, the elegant de Havilland D.H.91 Albatross, was unusual in having a plywood fuselage, which some commentators have since considered would have prevented it from competing effectively against the all-metal American aircraft, although the wartime de Havilland Mosquito fighter-bomber was none the worse for this form of construction. Entering Imperial Airways service in November 1938, the Albatross was used mainly on the European routes, to which it brought a useful 210 mph cruising speed and a generous 1,000 mile range. Normal passenger capacity was twenty-two.

Further development of the Albatross was cut short by the outbreak of World War II in 1939, at which time the prototype Fairey machine was under construction.

Less affected by World War II were the Junkers Ju.90 and the Focke-Wulf Fw.200 Condor, which were readily adaptable for military transport and maritime-reconnaissance duties. These forty-passenger aircraft were in service with Deutsche Lufthansa from 1938 onwards and could cruise at speeds well in excess of 200 mph. Possibly the faster aircraft was the Ju.90, with a top speed of 280 mph. Nevertheless, A. Henke was able to prove that the Fw.200 was also capable of a sparkling performance when, on 28–30 November 1938, he captained one on a flight from Berlin to Hanoi and Tokyo. It took $34\frac{1}{4}$ hours from Berlin to Hanoi, with an average speed of 151.23 mph, and $46\frac{1}{3}$ hours from Berlin to Tokyo, with an average speed of 119 mph. Earlier that same year on

13–14 August, he had flown from New York to Berlin in twenty hours at an average speed of 199.7 mph. Both aircraft types served throughout World War II with the Luftwaffe, but before the outbreak of war some airliner versions were exported, including one ordered by the Danish airline, D.D.L.

Not prepared to be beaten by the awakened Europeans, Douglas flew the prototype DC-4 four-engined airliner for the first time in June 1938, introducing the tricycle undercarriage and power-operated controls to airliner design for the first time. The aircraft did not go into production immediately, and the DC-4 which started to enter U.S.A.A.F. service in 1942 as the C-54, differed from the prototype which was a triple-fin machine with accommodation for forty passengers.

Further development of the DC-4 was to lead to the DC-6 and ultimately the long-range DC-7 of the post-war period.

Almost all these four-engined airliners were built in very small numbers, excepting the German aircraft, and although the DC-4 was ultimately built in large numbers, this was some time away and the aircraft was therefore a contemporary of such later types as the Lockheed Constellation and Avro York.

No exception to this trend for small-volume production came from Boeing, who built the most futuristic of all the immediate pre-war aircraft, in spite of its retaining a tailwheel. With a proven structural strength and safety, the Boeing 307 Stratoliner was the first pressurized airliner. The prototype first flew on 31 December 1938. Five production 307Bs entered service with Transcontinental and Western Air in April 1940 for service on the American domestic trunk routes. Only ten production aircraft were built for T.W.A. and Pan American World Airways, and after wartime service on charter to the United States Government the aircraft eventually passed to new owners. Their first major accident did not come until 1962 with a French charter airline.

All these new aircraft gave the passenger improved comfort. In many, such as the Fw.200, the Ju.90 and the Albatross, passengers sat two abreast on either side of the gangway in large fixed seats, with alternate rows facing backwards, railway-style. Others differed little from the airliner of today, and the de Havilland D.H.95 Flamingo, a twin-engined airliner for Imperial Airways which first flew in December 1938, actually had reclining seats.

Still with the emphasis on speed, Lockheed introduced and improved an up-rated version of the Electra I, the Lockheed 18 Lodestar in 1939, which saw service with a number of airlines and with the U.S.A.A.F. as the C-56.

World War II provided some justification for Sir Richard Fairey's belief that war hinders rather than assists aviation development, with many promising aircraft lost to civil aviation. But wartime development not only paved the way for another generation of airliners with improved performance; it also assisted in providing experience of long-haul transport operation, which was to stimulate airliner development further. In the meantime, the airlines and aircraft manufacturers could take pride in the fact that civil air speed had almost trebled in just twenty years, and reliability and passenger comfort had improved out of all recognition. An English judge was able to remark during a claim for damages before World War II that, 'today aeroplanes travel with the regularity of express trains'.

A lakeside scene in Finland does not provide the ideal photographic conditions for this 1924 Junkers F-13 monoplane airliner, and the large crowd of spectators is virtually unidentifiable. The real significance lies in the fact that even at such an early date the aeroplane had left the routes between the major capitals, where it could serve businessmen and diplomats, and was already helping the more lightly populated areas – in fact the aeroplane was more reliable throughout the year than the ferry, which was more at the mercy of the Baltic ice! Finnair operated only seaplanes for many years. (Photo: Finnair.)

Right: One of the first 'speed planes', the Lockheed Vega, with cruising speed depending on version ranging from 100 mph upwards to 135 mph; in its final form the maximum of 145 mph was possible. The plywood monocoque fuselage set a new standard in airframe construction, and the Vega was used on a number of pioneering flights, including some of those of Wiley Post and Amelia Earhart. (Photo: Lockheed Aircraft.)

Famous plane, famous passengers. The first Fokker Trimotor in the United States was for Lieutenant-Commander R. E. Byrd, u.s.n., who made the first flights over both North and South Poles in the aircraft. However, on this occasion the aircraft is carrying Anthony Fokker, and, amongst other passengers, Juan T. Trippe, founder of Pan American Airways, to Florida and Cuba in December 1925. (Photo: Smithsonian Institution, Washington.)

Looking almost like an overgrown fighter, the Lockheed Air Express of 1928 was the only parasol-wing design from the company. It brought a useful 150-mph cruising speed to air transport. It was once flown from Los Angeles to New York in just nineteen hours, including three stops and four passengers, and on another occasion flew non-stop coast-to-coast in eighteen-and-a-third hours. (Photo: Lockheed Aircraft.)

A rival to Fokker's predominance in the larger airliner market was the Ford Trimotor, a no less famous aircraft. This fine specimen is seen while being operated by Transcontinental Air Transport, which ran a rail-air service across the United States during the late 1920s and early 1930s. Charles Lindbergh was associated with the airline, a predecessor of T.W.A. (Photo: T.W.A.)

A Lockheed Orion in 1931, operating on the Los Angeles–San Francisco service of Varney Air Service, a predecessor of United Airlines. This was the ultimate in Lockheed's interpretation of the 'speed plane' concept. (Photo: Lockheed Aircraft.)

A development of the original F.VIIb, the Fokker F.XVIII was used on the longer services of K.L.M.-Royal Dutch Airlines, a company formed in 1919, initially using Aircraft Transport and Trading Aircos on a service to London. However, like Imperial Airways, K.L.M. had a massive colonial responsibility. This aircraft, PH-AIP 'Pelikaan', made a record return flight from Amsterdam to Djakarta between 18 and 30 December 1933, taking a total journey time of 201 hours, flying time 160 hours 20 minutes, for 17,650 miles. The weekly scheduled service took ten days, 81 hours flying time, each way. (Photo: K.L.M., Royal Dutch Airlines.)

This Douglas DC-2 surprised many by coming second in the October 1934 MacRobertson International Air Race from London to Melbourne and winning first prize in the handicap section. First was the de Havilland Comet racing aeroplane. The DC-2, PH-AJU 'Uiver', of K.L.M.-Royal Dutch Airlines, flew the 12,350 miles in 90 hours 17 minutes, out of which flying time accounted for just 71 hours 28 minutes. (Photo: K.L.M.-Royal Dutch Airlines.)

Opposite, top: Born 1935, still going strong. The Douglas DC-3 was a development of the earlier DC-1 and DC-2, with more than 13,000 aircraft having been built, mainly during World War II. Designed originally as a sleeper aircraft, the DC-3 usually ended up on ordinary passenger or military trooping operations, on which many of the occupants would have been surprised at the thought of anyone trying to sleep in such an aircraft! After World War II, the aircraft was used increasingly for freight duties as the demand for air freight rose and many DC-3s were retired from passenger operations. This Finnair aircraft is seen loading air freight during the mid-1960s. (Photo: Finnair.)

Opposite, middle: A contemporary of the Douglas DC-3, and a faster although less widely-used aircraft, was the Lockheed Model 10 Electra, seen here helping to give British Airways a more go-ahead image than the rival Imperial Airways. (Photo: Lockheed Aircraft.)

Opposite, below: Little explanation was required on this Short Brothers advertisement for the Empire flying-boat, built for the Empire Air Mail Scheme and operated by Imperial Airways and Qantas Empire Airways. A luxurious aircraft, and forerunner of the World War II Sunderland maritime-reconnaissance flying-boat, the graceful Empire could not compete with landplanes.

The dumpy Boeing 307 Stratoliner was built in small numbers for Pan American and T.W.A. before the outbreak of World War II, and was the first pressurized airliner to enter service. It deserved a better opportunity for widespread use, proving to be a long-lasting and reliable design. This is a Pan American example. (Photo: Boeing Aircraft.)

A cut-away view of a T.W.A. Boeing Stratoliner, showing just how the aircraft bridged the gap between the early airliners and the modern aircraft. Apart from pressurization, there are such other modern features as underfloor cargo stowage and the 'homing direction finder' based on radio beacons, along with such old-fashioned, if still desirable, features as berth accommodation – normally a flying-boat feature. (Photo: Boeing Aircraft.)

The ultimate in Lockheed's 1930s airliner development was the Model 18 Lodestar, arriving too late for the civil market in war-torn Europe but having military applications and still entering commercial service in the Americas and South Africa. It was still a relatively small aircraft, however, with only fourteen passenger seats but a speed not far short of 300 mph. (Photo: Lockheed Aircraft.)

Chapter 6

FIGHTER SPEEDS BETWEEN THE WARS

Confusion of types – overtaken by bombers – fighters close the speed gap.

To many, the most attractive aeroplane built, the Hawker Hart, of which this is a preserved example, brought bomber speeds to a par with those of the fighter of the late 1920s. A good example of Sir Sydney Camm's genius, the Hart was the start of a family of warplanes for every combat duty, including the Fury, the R.A.F.'s first 200 mph-plus fighter. (Photo: Hawker Siddeley Aviation.)

Although many of the aircraft manufacturers involved in building record-breaking and racing machines between the two wars also built fighter aircraft, it was rare for a variation of a fighter type to hold a record. Indeed, this situation did not occur until the appearance of the Messerschmitt Bf.109 in Germany in the late 1930s. This gap between fighter development and the development of racing and record aircraft did not reopen after World War II, but during the 1920s and for much of the 1930s the record aircraft and the fighter went their separate ways.

At first there was a calm after the storm. Plentiful wartime military aircraft supply followed by massive peacetime reductions in defence expenditure, meant that aircraft production dwindled after World War I. So did the development of new aircraft types. Many wartime designs remained in production and service for some years after the coming of peace – the D.H.9A and Bristol Fighter were but two examples – but in the end the inter-war period had to be one of development in military aviation, as in other fields of aeronautical activity. This was not without some relative neglect of the fighter, largely due to the value of the fighter-bomber, day bomber and general-purpose aircraft for the colonial police duties so important to the major powers. The paradoxical result of this trend was that on three occasions a new bomber type was developed capable of outpacing its own fighter 'support' as well as its opponents.

At a time when R.A.F. bombers were hard pressed to exceed 100 mph in level flight and fighters such as the Armstrong-Whitworth Siskin and Gloster Grebe managed a meagre 10 mph gain over the 125 mph maximum of the 1918 Bristol Fighter, Fairey built the Fox day bomber with a maximum speed of no less than 150 mph! A private venture, the Fox first flew in 1925 with a single 480-h.p. Curtiss D-12 engine. The aircraft was a shining example of what could be achieved once a good designer was given his freedom, and the contrast with aircraft built to rigid official specifications was all the more marked because of the prevalence of the latter.

The Chief of the Air Staff, Sir Hugh Trenchard, was so impressed with the performance of the aircraft that he immediately ordered a squadron. Alas, such were the financial constraints of the period that this was the one and only Fox squadron for the Royal Air Force.

Committee-designed aircraft have never been ideal because of the temptation to try and embody too many performance criteria in one type, so that it satisfies few needs well. This is the penalty of bureaucratic intervention in airframe design, and possibly a reason for the shortage of really good Soviet designs. It could also be the source of the saying that a camel is a horse designed by a committee. Equally apt, a pelican is an eagle designed by a committee, the talons having been traded for amphibious capability, no doubt!

Fairey's chief designer, Marcelle Lobelle, had been impressed by a visit to the United States during which he had seen the Curtiss racing and fighter aircraft. The Fox incorporated many of the Curtiss features, to the lasting benefit of British design.

Even so, the Gloster Gamecock fighter of 1925 was just capable of 150 mph, and the Boeing F2B and Curtiss Hawk were no faster than this. Boeing's P-12 and Bristol's Bulldog fighters of 1929 were faster, with a maximum speed of 175 mph, but the new Hawker Hart day bomber was capable of 184 mph – a higher speed than that of any contemporary fighter!

The truth of the matter really lay in the R.A.F.'s idea of a day bomber. Such aircraft were little more than fighter-bombers; this characteristic was underlined by the development of a family of fighter, fighter-bomber, liaison and ground-attack aircraft from the basic Hart.

It was a development of the Hart, the Hawker Fury fighter, which in 1931 became the first aircraft in the Royal Air Force to exceed 200 mph. This was a far from amazing turn of speed and just about half that of the Supermarine S.6B. But the phenomenon was accepted, just as in 1925 the Americans accepted Curtiss and Boeing fighters some 80-mph slower than the Schneider Trophy winner.

Boeing at this time was building the P-26, America's first monoplane fighter, with a maximum speed of 215 mph from its single 550-h.p. Pratt and Whitney radial engine. This was not too far ahead in terms of speed of the first modern medium bomber, the Boeing B-9, which could coax 185 mph from its two 575-h.p. Pratt and Whitney engines.

These were all left behind by the Martin B-10 when it appeared in 1932. The Martin's speed of 265 mph could not be easily ignored, because this aircraft, faster than any fighter, was a true bomber. An all-metal monoplane, the twin-engined B-10 set the pattern for bomber development for more than a decade. Against this type of bomber development the production of fighter aircraft, such as the Gloster Gauntlet biplane of 1934 with a maximum speed of around 230 mph, was hardly adequate, even allowing for any ground attack or fighter-bomber capability.

However, a more proper order of military aircraft speeds was restored during late 1935 and 1936 with the advent of the Boeing B-17 heavy bomber (via the ill-fated Boeing 299 prototype), and the Messerschmitt Bf.109 and Hawker Hurricane fighters. While heavy-bomber speeds increased to 280 mph, those of the fighters climbed over the 300-mph mark: early Hurricanes were capable of 315 mph in level flight.

Both the Hurricane and the Bf.109 were able to improve upon their initial speeds, to the extent of finally closing the gap between the fighter and the

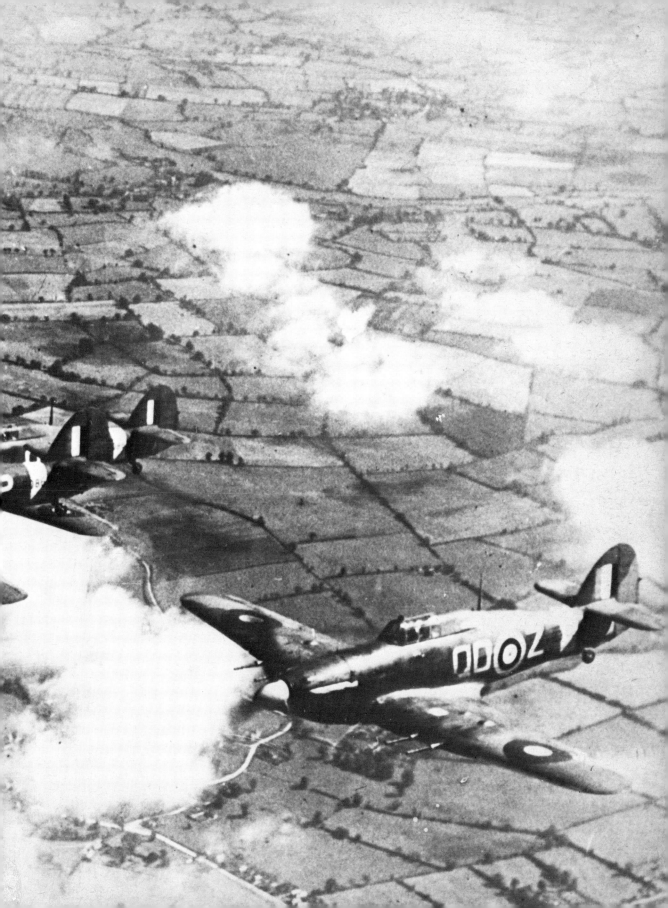

An aircraft of unusual design, the Bell P-39 Airacobra had the engine mounted behind the pilot's cockpit. It was not a fast aircraft by the standards of 1939, but it did well as a tank buster for the Russians. Note the cannon mounted in the propeller hub. (Photo: Bell Aerospace.)

Opposite: A classical view of the Supermarine Spitfire fighter, which was entering R.A.F. service shortly before the outbreak of World War II in 1939. In a dive, Spitfires were the fastest propeller aircraft ever built; Squadron Leader Martindale, R.A.F., took a Spitfire XI to 620 mph in 1944. (Photo: British Aircraft Corporation.)

When Boeing introduced its 299 bomber prototype in 1935, the balance of speed and defensive power in military aircraft development seemed ready to swing back to the bomber once again, until the prototype crashed. However, not only was the B-17 Fortress ready for America's entry into World War II, but this development of the 299 also had to be built by other manufacturers in order to keep pace with demand. This example was by Lockheed. (Photo: Lockheed Aircraft.)

'Bristol' BLENHEIM BOMBER

THE BRISTOL AEROPLANE C° L^{TD} FILTON BRISTOL

Bristol produced a family of night fighters and light bombers in the Beaufighter, Beaufort and Blenheim series, capable of speeds of around 280–300 mph and entering R.A.F. service just before and after the start of World War II.

Designed to meet a Royal Air Force specification for a light bomber and maritime-reconnaissance aircraft, the Hudson was an adaptation of the Model 14 civil design and also served as a V.I.P. transport. (Photo: Lockheed Aircraft.)

First flown in 1939, the twin-boom,
twin-engined Lockheed P-38
Lightning was soon in service with
the R.A.F. and U.S.A.A.F. It
provided a much-needed long-
range escort fighter capability, and
was the first 400 mph-plus
operational aircraft. (Photo:
Lockheed Aircraft.)

France's best fighter at the
outbreak of World War II was the
Dewoitine D.520, available in very
small numbers and capable of a
maximum speed of around 380
mph. The fault lay more with the
politicians than with the designers
or the industry. (Photo: Musée de
l'Air, Paris.)

Backbone of the Armée de l'Air at
the outbreak of World War II was
the Morane-Saulnier MS.406, and
notwithstanding its modern design
this aircraft was slower than those
of the Luftwaffe, with a speed in
the region of 320 mph. (Photo:
Musée de l'Air, Paris.)

Chapter 7

WORLD WAR II

Official records suspended again – the peak of piston-engined performance – the turbojet era, rocket-powered aircraft.

The first monoplane fighter, both for the R.A.F. and for Sir Sydney Camm, the Hawker Hurricane first flew in 1935 and played the major role in the Battle of Britain in 1940. Outpaced by the later Supermarine Spitfire, and the German opposition, the sturdy Hurricane could still counter German bombers. (Photo: Richard E. Gardner.)

World War II differed from World War I in many respects. If aviation was important during World War I, it was many times more so during World War II. The latter was lost and won in the air, although neither defeat nor victory could have been complete had these not been shared by land and sea forces. Success at sea and on land was allied to the skilful development of covering air power, and Germany's failure to master this technique contributed greatly to her defeat.

Whereas World War I did not see technological advance in the air but merely a more rapid and widespread application of existing technology and its use to the full, World War II saw technological advance of far reaching importance not only in aviation, but also in the related field of rocketry. A number of civil projects were denied their rightful place in the order of things because of the onset of the war, but they were made obsolete by the progress made between 1939 and 1945.

Most important from the aspect of speed in the air, World War II put the record-breaking aircraft of the pre-war period and their developments in frontline service. The Supermarine S.6B had given birth to the Supermarine Spitfire fighter. The Spitfire was soon locked in combat, first over France and then over England, with the Messerschmitt Bf.109 fighter, a version of which, the Bf.109R, had established the last pre-war speed records. Italy also gained some advantage from her racing and record-breaking aircraft of the 1920s and 1930s, although Italian excellence in airframe design was not matched by their engine-design, and some Italian aircraft did not fulfil their true potential until re-engined with German powerplants.

In one sense both wars were similar. Official speed records came to an end with the outbreak of hostilities, for reasons of secrecy. In wartime it is only possible to identify the fastest aircraft types.

It could be said that the record set by Flugkapitan Fritz Wendel of the Luftwaffe (469.22 mph on 26 April 1939) in a specially modified Messerschmitt Bf.109R stood until 1945, although in the meantime the speed of operational aircraft came to exceed that of the record holder. The Bf.109's speed rose from the 323 mph of the original operational aircraft used during the Spanish Civil War, which had a 1,000-h.p. Daimler-Benz 600AL engine, to the 428 mph of the Bf.109G-10 version of 1944, which used a 1,800-h.p. Daimler-Benz 605D engine.

A successful design in many respects, of which a record 33,000 aircraft were built between 1939 and 1945, the Bf.109 suffered from drawbacks in operational service which made it inferior to the slightly slower Spitfire. One of these was the weakness of the tail structure compared with that of the sturdier British aircraft. Another was the lack of effective armour-plating to help minimize injury to the pilot. An advantage of the Bf.109 was the provision – on all but the earliest aircraft – of a machine gun firing through the propeller boss.

A less numerous but very effective counterpart of the Bf.109 was the Focke-Wulf Fw.190, designed by Professor Willi Messerschmitt's main contemporary, Kurt Tank. After a first flight in 1939 production models of the Fw.190, with its fourteen cylinder 1,600-h.p. B.M.W. 801C radial engine, entered Luftwaffe service in late 1940, and some 20,000 were built before the end of the war. Increased power and reductions in airframe weight pushed the maximum speed at altitude from 370 mph on the prototypes to more than 420 mph during 1943 with the D-9 version using a 2,240-h.p. Jumo 213A-1 engine. The fastest aircraft of this type came with the developed Ta.152 of 1945, of which a high altitude version was capable of 472 mph in level flight and a maximum altitude of 41,000 feet.

Against this, the Supermarine Spitfire in 1940 had a maximum speed in level flight of 350 mph, considerably less than the 400 mph plus of the Supermarine S.6B seaplane of 1931. However, as World War I had amply illustrated, speed is not the sole criteria for a warplane, although it is very important. Attention must also be paid to handling qualities, lack of temperament, range and war-load. In its superior manoeuvrability, more rugged construction and better protection for the pilot, the Spitfire was a good match for the Bf.109. Ultimately more than 23,000 Spitfires were built, mainly with Rolls-Royce Merlin engines, although many of the later versions, numbering about 3,000, used the Rolls-Royce Griffon, with a horse-power rating often in excess of 2,000, and a number of aircraft used Merlin engines built under licence in the United States by Packard.

Speed improved considerably over the Spitfire's production life, which lasted until the end of World War II. The Griffon-powered Mark XIV of 1944 managed 439 mph at 24,500 feet, and later versions still achieved 446 mph. It is generally accepted that, by the end of World War II, some production Spitfires frequently could, and did, exceed 600 mph in dives; in 1944, Squadron Leader Martindale, R.A.F., dived a Spitfire XI to 620 mph!

The gain in speed was accompanied by many other improvements. The eight small-calibre machine guns of early Spitfires were replaced by heavier calibre weapons, usually known as cannon, of which the aircraft carried four; sometimes there were rockets under the wings for anti-train or tank-busting duties. Other aircraft of the period incorporated similar improvements, so that after two or three years in production the difference in performance of any given aircraft type tended to be considerable. Airframe modifications, including the cut-away fuselages aft of the cockpit on certain versions of the Spitfire, Thunderbolt and Mustang, gave visual effect to these developments.

World War II also marked a certain reversing of the trend towards specialization of aircraft design which had begun in 1916. This reversal has survived to the present day and is a direct result of progress on the powerplant and airframe fronts which enables one basic design to be adaptable for a variety of rôles, and to excel at many of them.

The Spitfire was outnumbered during the Battle of Britain by the Hawker Hurricane, which had had the distinction of being the Royal Air Force's first monoplane fighter design. The Hurricane could not be regarded as the fastest aircraft of World War II by any means, but it was a dependable and potent mount of some of the R.A.F.'s best pilots. Early production versions, in service with the Royal Air Force during 1937, were capable of 316 mph, but it is doubtful whether even the later marks of Hurricane could manage much more than 400 mph in a dive.

That an aircraft such as the Hurricane should have fared relatively well against the Messerschmitt Bf.109 is not without significance. It must be remembered that the Royal Air Force pilot of 1940 was usually a young and inexperienced aviator, whereas his German opponent had usually been blooded during service with the Legion Condor during the Spanish Civil War. One can only guess at the reasons for the outcome, but one of them must be the fact that a hard-pressed defender fights harder than an attacker. Moreover, a soldier whose country is being invaded is tempted to desert and protect his family, an airman knows that he cannot do better or be more effective than when he is in the air fighting.

Range limitations must also have prompted the Germans to try to break off combat earlier. This gave the British yet another advantage. The point is that neither an advantage in speed nor in numbers was sufficient to give Germany victory in the Battle of Britain, and this underlines the importance of qualities other than speed in aircraft design.

Certainly, a production of 13,000 Hurricanes was justified.

The Japanese Mitsubishi A6M1 Reisen – popularly known as the Zero – was another leading aircraft of the war years which was not in the forefront of aviation development; it was probably effective for as long as it was only because of the original tendency for the Allies to retain their best equipment for the European theatre. The first production versions in 1939 had a maximum speed of just 304 mph, and the fastest mark in 1946, with a 1,560-h.p. Mitsubishi Kinsei 62 engine, could manage only 356 mph. Only 10,000 Zeros were built, including the 'throw away' Kamikaze (suicide) versions on which the pilot was also not re-usable!

Apparently, a failing of many Japanese aircraft, including the Zero, was the absence of a powerplant of high performance.

The Soviet aircraft industry at this time was unable to produce any aircraft with a performance to match those of the Axis or the Allies, although after changing sides the Soviet Union was able to benefit from British and American know-how and hardware.

A disastrous policy of nationalization and re-organization had left the French aircraft industry in disarray by 1939, with few modern aircraft available to the Armée de l'Air for the defence of France, let alone extensive African and Far Eastern possessions. The few surviving private enterprise firms did their best, although no doubt the threat of nationalization must have hindered investment. France's best fighter in 1940, the Dewoitine D.520, was available in very small numbers. This left the backbone of the nation's defence to a slower aircraft, the Morane-Saulnier MS.406 fighter. The promising Breguet Br.690 was only just entering service; this was a twin-engined aircraft better suited for long-range escort duties, or, had radar been available for fighter defence in France, night-fighter duties. Only eighty of the Bloch M.B.151 fighter monoplanes, with a 320-mph maximum speed, were in service in 1939.

A policy of pacifism and defeatism in the Netherlands between the wars had prevented Fokker from turning that company's undoubted talents towards the development of advanced military aircraft. The result was that in 1939 the main fighter of the Royal Netherlands Army Air Service, the Fokker D.23, a twin-engined tractor and pusher propeller design with a twin-boom airframe, was not available in adequate numbers; nor was it really a match for the Bf.109. A good Dutch fighter of the late 1930s, the Fokker D.21, entered service in small numbers only.

Italy's Reggiane Re.2000 Falco monoplane fighter suffered from structural weakness, although the Re.2001 Falco II of 1941 had this failing remedied; it also had a 1,175-h.p. water-cooled in-line engine instead of the radial engine of the earlier aircraft which had given rise to drag. The improved aircraft was capable of a maximum speed of 360 mph.

As usual, the maximum speeds were those of the fighters. Nevertheless, significant increases in speed had also been achieved by other aircraft types, including the new long-range escort fighter and night fighter, which at the outset of the war were represented by the British Bristol Beaufighter and the German Messerschmitt Me.110; both were twin-engined aircraft capable of speeds in the region of 325 mph. Next down the speed list came the medium bombers, the Bristol Blenheim and Beaufort and the German Junkers Ju.88, with speeds of just over 300 mph. As the war progressed, there tended to be some merging of the long-range fighter or night fighter types with the lighter medium bombers.

Obviously, the heavy bombers which proved to be such a potent weapon during the war could put up little pretence at setting speed records. As the war progressed, bomber speeds did rise – from approximately 240 to 360 mph.

One of the most notable aircraft of the war – and the one which did most to expand the rôle of a particular aircraft type – was the de Havilland D.H.98 Mosquito. This twin-engined mid-wing monoplane was intended at the time of the prototype's maiden flight on 25 November 1940 to be a light-medium bomber. The feature of the Mosquito most frequently noted was the construction of the airframe almost entirely of wood.

The early 1941 production versions of the Mosquito had a maximum speed of 350 mph from two 1,250-h.p. Rolls-Royce Merlin 21 piston engines. These aircraft were soon modified to act as night fighters, long-range escort fighters, fighter-bombers and photographic-reconnaissance aircraft. The ability to carry up to 4,000 lb. of bombs marked the Mosquito as a good medium bomber; but the ability to operate over a radius of 750 miles with this load, even farther as an escort fighter, and an ultimate maximum speed in level flight of more than 400 mph, showed the aircraft's versatility of performance.

Total production of the Mosquito in the United Kingdom, Canada and Australia reached 8,000 aircraft: a notable figure for such a large aeroplane.

In one respect the Mosquito could also be said to have advanced civil air speeds dramatically. Unarmed 'civil' versions of the aircraft were used by the new British Overseas Airways Corporation during the war on a Scotland–Sweden service. This particular air route passed over occupied Norway, hence the need for an aircraft with rather more speed than the usual transport type. As Sweden was a neutral country the aircraft had to be unarmed and civilian operated. The purposes of the exercise were to fly V.I.P.s and diplomats in both directions and to carry much-needed ball bearings from Sweden to Scotland.

Passenger accommodation in the Mosquito bore little resemblance to pre-war standards or even the earliest standards. It consisted of a stretcher suspended in the bomb bay over the bomb doors! Passengers spent the flight lying on the stretcher. Reputedly, particularly obnoxious passengers would find the bomb doors opening beneath their stretcher while flying over the North Sea! Stewardess service was not provided on these flights! It was some years before the Mosquito's 400 mph was exceeded by an airliner.

The American aircraft industry had, during the 1930s, produced a number of successful military designs in addition to its civil successes and in spite of a restricted defence budget. Aircraft such as the Boeing B-9 and Martin B-10 bombers not only paved the way for the modern bomber; being faster than most contemporary fighter or pursuit, aircraft, they spurred fighter development.

G

At the time of America's entry into World War II in 1941, the U.S.A.A.F.'s two main fighter types were the Bell P-39 Airacobra and the slower Curtiss P-40 Warhawk. The Airacobra had a maximum speed of around 360 mph and the unusual feature of an engine mounted behind the cockpit, even though the single tractor propeller was mounted in the usual place in front of the cockpit. The Airacobra also had a Bf.109-type propeller-boss machine gun and a tricycle undercarriage; originally intended as an interceptor, this aircraft proved to be a successful ground-attack type.

These aircraft were already being joined by two new American aircraft of outstanding performance and durability: the North American P-51 Mustang and the Republic P-47 Thunderbolt. The Mustang had first flown in October 1940, as the Apache, using a 1,100-h.p. Allison V-1710 engine, whereas the Thunderbolt did not fly until May 1941, when it used a 2,000-h.p. Pratt and Whitney XR-2800-21 radial engine. The aircraft differed in that the Mustang was essentially a long-range escort fighter and the Thunderbolt tended towards the fighter-bomber rôle. They entered service in 1942 with the Royal Air Force and United States Army Air Force, respectively, although the Mustang also entered U.S.A.A.F. service in due course.

Initially, both the Mustang and the Thunderbolt had maximum speeds of around 400 mph, although the performance of the Mustang improved substantially with the substitution of a Packard-built Rolls-Royce Merlin of 1,520 h.p. for the original Allison powerplant, giving a speed nearer 440 mph. The fastest production Mustang, the U.S.A.A.F.'s P-51H, had a maximum speed of 487 mph, and the specially-modified P-51M had 491 mph claimed for it. A lightened airframe, engine modifications and additional streamlining to the engine cowling enabled the fastest Thunderbolt, the XP-47J, of which only one was built, to reach 504 mph. Both aircraft remained in service with Latin American air forces until the early 1970s.

In two years of war aircraft speeds rose from a maximum of 350–380 mph to 440 mph. Even the Soviet Union was able to squeeze 400 mph out of its best fighter design at the time – the Mikoyan-Gurevich MiG-3 fighter of 1941.

The leading air aces were already well established. Group Captain 'Johnie' Johnson of the R.A.F., the leading British ace, flew a Supermarine Spitfire; the leading German ace, Major Ernst Hartmann, with a total of 352 confirmed victories, gained mainly on the Russian front, flew a Messerschmitt Bf.109G. Germany's main ace in the West, Hauptman Hans-Joachim Marseille (with 158 confirmed victories) flew a Messerschmitt Bf.109E. It should be remembered that the high totals of the Luftwaffe pilots included operations during the Spanish Civil War and then against countries such as Poland, whose obsolescent aircraft attempted to stem an overwhelming German tide.

The disadvantage of the piston-engined aircraft was the inefficiency of its propeller, in spite of the progress made during the late 1920s and 1930s which had resulted in most aircraft being equipped with variable-pitch propellers allowing the optimum pitch settings for take-off and for flight. As speed increased beyond 500 mph, propeller efficiency fell off rapidly, and much the same happened as altitude exceeded 25,000 feet. It was an opportune moment for a system of propulsion which was at its most effective above the point where the propeller failed to give extra performance.

Work on jet engines had taken place in Germany and the United Kingdom between the wars; Dr Hans von Ohain and Frank Whittle (later Air Commodore Sir Frank Whittle) worked independently. The prototype machines established no speed records, however. The Heinkel He.178 first flew in Germany in 1939, with a maximum speed of 435 mph, which was a good performance for a test-bed. This was followed by the Caproni-Campini in Italy, which flew in 1940 and 1941, but seems to have had a maximum speed of just 250 mph. The Italian aircraft was an oddity by any standard. A ducted fan was driven by a piston engine, and fuel injected into the resulting airstream was ignited. Britain's Gloster E.28/39 first flew on 15 May 1941.

Operational jet aircraft followed, although their entry into service with the Luftwaffe and Royal Air Force proved to be more leisurely than might have been expected. The Messerschmitt Me.262A-1a, the first operational jet aircraft, first flew in 1942. It was followed a year later by the Gloster Meteor, but neither of these twin-engined monoplanes entered service until 1944. The German aircraft was the faster; it had a maximum speed of 525 mph from its Junkers Jumo 109-004B-1 axial-flow engines, whereas the developed Whittle centrifugal-flow engines on the Meteor gave a maximum speed of 480 mph on the early production versions. On the other hand, the British aircraft had the advantage of continuous development, so that by the end of the war it was capable of a maximum speed in level flight of more than 550 mph. It went on to set the post-war air speed records late in 1945.

Other aircraft followed these early jets. In Germany, the Heinkel He.162 Salamander, or 'Volksjager', designed in six months and with a maximum speed of 520 mph, entered service and soon earned a reputation for being un-safe. The first jet bomber (the Junkers Ju.287) flew but did not enter service, although the Arado Ar.234 jet reconnaissance aircraft was used as a bomber. In the United Kingdom, the de Havilland Vampire, a single-engined twin-boom fighter design, first flew in 1943, but did not enter R.A.F. service until after World War II.

American interest in the new form of propulsion was lukewarm, possibly because of the importance placed by the Americans on range and the inability of the early jet to match the endurance of the piston-engined fighter. General Electric did build Whittle turbojets under licence so that Bell could produce all of fourteen Airacomet twin-jet aircraft. They first flew in October 1942 but did not enter service. The first operational American jet aircraft was the Lockheed P-80, which did not enter production until after the war.

The Japanese produced some jet-powered Kamikaze aircraft, but these (and rocket-powered variants) were not particularly fast. Perhaps they were considered cheaper for suicide operations than even 'throw away' variants of aircraft like the Zero.

Undoubtedly the fastest aircraft of World War II was the Messerschmitt Me.163 Komet interceptor, a swept-wing aircraft first flown in 1944. Powered by a Walter liquid-fuel rocket motor, this, the first operational rocket-powered aircraft, had a maximum speed of 590 mph, coupled with a phenomenal rate of climb for the time. The pilot would glide the Komet back to base after being tow-launched by a bomber or ramp-launched from the ground. The effectiveness of the Komet against Allied bombers was severely reduced by poor endurance. A trials Komet actually exceeded 610 mph on one occasion.

The poor endurance of the early jet fighters and the even worse endurance of the rocket-powered interceptor made further development of the piston-engined aircraft both worthwhile and necessary. A good pilot could prove a piston-engined aircraft a fair match for many jets. During the Korean War Fairey Fireflies and Hawker Sea Furies were able to shoot down Mikoyan-Gurevich MiG-15 jet fighters.

Long-range escort fighters such as the P-38 Lightning of 1941 were capable of speeds in excess of 400 mph, and other new aircraft with high speeds included the Hawker Tempest, developed from the earlier Typhoon, and the Grumman F8F Bearcat. A single 2,000-h.p. Napier Sabre 'H' type engine powered the Tempest, while the Bearcat used a 2,500-h.p. Pratt and Whitney radial. Later Tempests used developed Sabre engines of up to 2,825 h.p., giving a speed in level flight of well over 450 mph.

An indication of the Tempest's performance is that the aircraft was the only type able to destroy the German V-1 jet-propelled flying-bombs, launched against British cities towards the end of World War II. At first the Tempest pilots destroyed the V-1s by flying alongside and turning them over with their wings; this caused a sufficient upset for the V-1 to crash. The practice had soon to be discouraged in favour of more conventional methods of destruction after a number of 'accidents' in which V-1s blew up immediately on being toppled off course.

Although the Grumman Bearcat was a match for the Tempest in terms of speed, both aircraft were just behind the two fastest piston-engined aircraft to enter operational service, although neither of the latter flew during World War II for other than test purposes. The de Havilland Hornet long-range fighter, originally intended for service in the Pacific against Japan, was able to reach some 480 mph in level flight with two Rolls-Royce Merlin engines, and the Hawker Sea Fury (which first flew in September 1944 but did not enter Fleet Air Arm service until 1947) had a marginally higher speed.

Although there was certainly a six-year pause in speed record attempts, the fact remains that by the end of World War II, as in World War I, operational aircraft with a high fuel load, armed and ready for battle, were able to exceed the record speeds of specially modified pre-war aircraft. The speeds of these operational aircraft had also risen during the period by more than 200 mph. Yet, just as the rotary engine reached its limitations in 1918, the radial and in-line engine followed suit in 1945, requiring a far more drastic change, from piston to turbine power.

Below: One of the most important American fighter aircraft of World War II, the Republic P-47 Thunderbolt had a maximum speed of around 400 mph in the standard version, shown here before take-off. Later models had a blister canopy and cut-down rear fuselage – a change from the humpback fashion of the early war period also shared by the Spitfire and Mustang. (Photo: Fairchild Hiller, Republic Division.)

A Messerschmitt Bf.109 in R.A.F. markings! Obviously a captured example. More than 30,000 Bf.109s were built, mainly during World War II. The prop-boss cannon can be seen clearly in this picture. (Photo: Smithsonian Institution, Washington.)

Fastest Thunderbolt and an unofficial speed record holder, the Republic XP-47J incorporated a lightened airframe, engine and aerodynamic modifications which enabled it to fly at 504 mph on 4 August 1944. It is open to question whether this aircraft or the Messerschmitt Me.262 was the first to fly faster than 500 mph in level flight, but certainly the XP-47J was the first propeller aircraft to do so, and it could probably maintain the speed for longer than the turbojet. However, the question is academic in the absence of an official record. (Photo: Fairchild Hiller, Republic Division.)

One of the most versatile aircraft of all time, the de Havilland Mosquito was originally designed as a bomber, but proved itself a good night fighter and reconnaissance aircraft, not to mention fast transport for B.O.A.C.! Well liked by those who knew it, the over-cautious or over-confident pilot could soon find that not all of the bites were reserved for the enemy! (Photo: Flight International.)

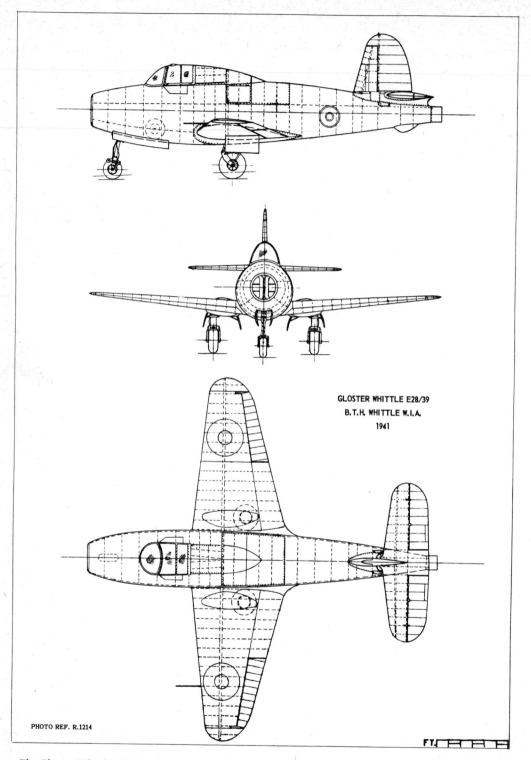

GLOSTER WHITTLE E28/39
B.T.H. WHITTLE W.I.A.
1941

PHOTO REF. R.1214

The Gloster Whittle E28/39 was the first British jet-propelled aircraft, although it could scarcely match, let alone exceed piston-engined aircraft speeds in 1941. It was a small aeroplane of conventional design and layout, as can be seen from this three-view general arrangement drawing. (Hawker Siddeley Aviation.)

The first operational jet aircraft, designed as a fighter by Willy Messerschmitt but pressed into service as a bomber on Hitler's insistence, the Messerschmitt Me.262, of which an A-1a version is shown here, entered Luftwaffe service in 1944. (Photo: Messerschmitt Archive.)

The fastest aircraft of World War II, and forerunner of both the post-war tail-less and air-launched designs, the Messerschmitt Me.163 Komet was designed to counter high-flying Allied bombers. It is seen here on its launching trolley. This aircraft is preserved in the Deutsches Museum, Munich. (Photo: Messerschmitt Archive.)

America's only World War II jet aircraft was the Bell P-59A Airacomet with American-built Whittle jet engines. Although designed as a fighter, it did not enter production and the numerous prototypes were used for research and test flying prior to the development of the Lockheed P-80 Shooting Star. (Photo: Bell Aerospace.)

Speeds of more than 400 mph were possible with the Hawker Typhoon, seen here in prototype form, although the aircraft was used mainly on ground-attack duties, gaining considerable fame and notoriety as a tank and train buster. (Photo: Hawker Siddeley Aviation.)

A development of the Typhoon, the Hawker Tempest and its near relative, the Sea Fury, represented the peak of piston-engined fighter development. They were capable of speeds in level flight of not much less than 500 mph, and were rugged and manoeuvrable aircraft as well. The Tempest gained fame for shooting down German V-1 flying-bombs towards the end of World War II. Some pilots preferred to fly alongside the V-1 and gently nudge it with a wingtip, throwing it off balance, but the R.A.F. had to stop this dangerous practice. (Photo: Hawker Siddeley Aviation.)

Chapter 8

RECORDS AFTER WORLD WAR II

First turbojet records – through the sound barrier, Anglo-
American rivalry – the Soviet challenge.

Britain's last official record holder, the Fairey Delta 2 research aircraft in flight. Proportionately, this
aircraft achieved the biggest ever jump forward in the speed record. It survives today in modified form
as the B.A.C. 221 research aircraft for the Concorde test programme. (Photo: Westland Aircraft.)

In marked contrast to the immediate post-World War I period, there was little delay after the end of World War II before first attempts were being made on the air speed record. No doubt much of this enthusiasm was born of impatience to put the new technology of the war period, the turbojet, to the test. It was not only vitally important to discover just how far and how quickly the new power unit could be developed; the aerodynamics of the aircraft also required investigation. More than ever before the speed record represented a step forward into the unknown.

Some importance was attributed to the need to raise the record past the significant 1,000 kmph mark, but the real challenge was that of the sound barrier.

The whole atmosphere of the record attempts had changed. Now, more than ever before, it was a deadly serious business. Gone were the fashionable and exciting sporting events with an attack on the speed record as a highspot. The days of the genuine company attempt were also gone, although often company pilots would fly aircraft developed under a government research contract or as the prototype of a new fighter aircraft. Some of the aircraft involved were research types, but for the most part it was the fighter which was allowed to set the pace, sometimes with only minor modifications for a record attempt. The days of the racing aeroplane had passed as far as very high speed flight was concerned.

Public interest had faded with the changes. Only a really spectacular achievement could catch the imagination or reach the headlines.

Naturally enough, the vehicle for the first post-war attempt was the Gloster Meteor IV, a cleaned-up development of the Meteor III, which was itself a development of the basic aircraft which had been the world's second jet fighter and the only Allied jet fighter to see service during World War II. The war had, of course, prevented any attempt from being made on the record, although the original Meteor could not have improved on the Bf.109R's record without modification: its maximum speed was 460 mph at sea level and about 475 mph at altitude; although the latter figure could at this time be discounted for record purposes.

After a number of preliminary tests on Meteor IIIs, two Meteor IV aircraft were provided for the record attempt – one, painted bright yellow, belonging to the manufacturer, Gloster Aeroplane, and the other, in standard Royal Air Force camouflage, belonging to the R.A.F. and carrying the name, 'Britannia'. The record attempt Meteor IVs differed from their operational counterparts only in details, such as longer nacelles for the two Rolls-Royce Derwent turbojets and some strengthening of the wings for the stresses of low-altitude high-speed flight. A flight in one of these aircraft at Moreton Valence on 19

October 1945 enabled Squadron Leader P. Stanbury, D.F.C., R.A.F., to attain an unofficial 603.00 mph.

The speed record unit then moved to Kent for trials over the record course in Herne Bay. It was at this time that the weather, which had been excellent throughout the autumn, changed for the worse. In fact, the highly seasonable November weather produced every kind of meteorological condition except those needed for an attempt on the record. The course lay roughly east–west, between turning points at the Isle of Sheppey in the east and the North Foreland in the west. Too strong a wind could cause problems of drift, and visibility of at least five, and preferably ten, miles was needed.

Some good performances were recorded by the two pilots during practice attempts. It had already been decided that 'Britannia' would be flown by Group Captain H. J. Wilson, A.F.C., R.A.F., the Commanding Officer of the Empire Test Pilots School at Cranfield, and the yellow Meteor would be flown by the Chief Test Pilot of Gloster Aeroplane, Eric Greenwood. On one occasion, Wilson flew at 590 mph at altitude and then, coming down to low level, made one flight at 601 mph and another at 609 mph over the course. Greenwood managed three runs over the course at an average speed of 525 mph until rapidly worsening visibility on the final run forced him to seek refuge in high altitude, and hastily!

Finally, on a dull and overcast Wednesday, 7 November 1945, the decision was taken to press ahead with an attempt on the record, substituting this for another morning's trial flying. In visibility of between seven and twelve miles, and with a twelve-knot wind blowing from East Anglia, Wilson flew over the course at 604 mph, then 608 mph, followed by 602 mph, and, after nearly abandoning the record attempt because of drift and difficulty in picking up the marker flares, finally at 611 mph. His average of 606.38 mph (975.67 kmph) was a new record. His fastest times were west–east with the wind.

Greenwood followed immediately, but could only manage 599 mph, 608 mph, 598 mph and 607 mph, although, to be fair, his performance must have been adversely affected by the strain of starter trouble, which had him in and out of his cockpit no less than four times before the fault could be rectified and he could take-off.

A further attempt using the Meteor was made in 1946 by the R.A.F.'s High Speed Flight. Although the same mark of aircraft was used, the pilots and the site of the course differed, with the measured three kilometres running part of the way between Littlehampton and Worthing on the Sussex Coast. An earlier start was made, with the Flight's Meteors ready for trials in the middle of August, but a typically English summer and technical problems resulted in three weeks of frustrating delays and few promising flights.

It was not until 7 September that, on a grey and wet evening, the Officer Commanding the High Speed Flight, Group Captain E. M. 'Teddy' Donaldson, D.S.O., A.F.C., R.A.F., was able to fly along the course five times, with an average speed for his four fastest runs of 615.78 mph, setting yet another record for the Meteor. Donaldson had a markedly bumpy ride during his flight, and as the conditions deteriorated further, Squadron Leader Waterton, who followed, could only manage an average of 614 mph.

The question now was just who would set the first official record of more than 1,000 kmph? Donaldson had managed an unofficial record of 1,003.31 kmph, or 623.45 mph, on his fastest run, but it was known that the Americans were getting ready to re-enter the speed record field. They were the only likely challengers; the Soviet Union was still struggling to match Western standards, and Europe was recovering from the effects of war.

In fact, the American challenge was a little longer in coming than either the Americans or the British expected. A number of problems concerning the P-80 Shooting Star confronted the aircraft's manufacturer, Lockheed. But the firm persisted with the aircraft, confidently expecting success in the end for what would, after all, be America's first operational jet fighter. This confidence paid off, for, on 19 June 1947, flying over the Muroc Dry Lake in California's Mojave Desert, Colonel Albert Boyd, U.S.A.A.F., flew a specially-modified P-80 to a new record of 623.8 mph. The record was the average of four runs at 617.1 mph, 614.7 mph, 632.5 mph and 630.5 mph.

Modifications to the aircraft used on the record-breaking flight included water-methanol injection and after-burning on the single Allison turbojet, and air intake extensions. The modifications accounted for many of the difficulties encountered during the attempt on the record, and the standard aircraft, and its widely-used trainer variant, was a reliable machine. One problem which Boyd did not have to contend with was that of poor visibility. His high speed flight was made in ideal weather conditions, and his course was marked by a twelve-foot wide black strip on the white desert surface and smoke flares at either end.

British feelings that the Meteor IV could have managed at least as good a performance under similar conditions did not prevent them from giving their American rivals full credit for a noteworthy achievement on what was recognized to be a fine aircraft.

In order to regain the record many Britons looked to the de Havilland DH.108 Swallow, a tailless research aircraft. This was known to be faster than any aircraft which the Americans could use for a record attempt, because their Bell XS-1, which was the first aircraft through the sound barrier in October 1947, was launched from a Superfortress bomber at altitude and therefore ineligible for a record attempt. The DH.108 was the first British aircraft through the sound barrier, on 6 September 1948. This event occurred after extensive testing, but it was not really premeditated. Test pilot John Derry was flying the aircraft from Hatfield when, while between Windsor and Aldershot at 40,000 feet on a fine morning, he felt that the right time had come and pushed the control column forward. The DH.108 slipped through the sound barrier without any buffeting or vibration during its dive. In spite of this performance, a number of accidents with the DH.108 and other tailless aircraft persuaded the British Government to abandon the project before any attempt on the absolute speed record could be made, although not before the aircraft managed to put a 100 kilometre closed circuit speed record to its credit.

In the meantime, the Americans were hanging on to the record in a convincing manner, aided by a revival of the old inter-war rivalry between the United States Navy and the United States Army, or United States Army Air Force as it had become. Within two months of the Shooting Star taking the record the United States Navy was ready with the Douglas D-558 Skystreak for a record attempt over the Muroc Dry Lake.

The Skystreak set two records in succession. The first was on 21 August 1947, when Commander Turner F. Caldwell, U.S.N., took the record to 640.74 mph. Just five days later Major Marion E. Carl, U.S.M.C., flew the aircraft at what many then considered to be the 'truly remarkable' speed of 650.92 mph. The real supremacy of the aircraft lay in the fact that it had not been modified in any way for the record attempt, and its General Electric TG-180 engine did not use special fuel, water injection or reheat.

The United States Air Force, which had come into existence by this time as a completely separate service, continued the distinguished and fruitful tradition of inter-service rivalry started on its behalf by the United States Army Air Corps and then the U.S.A.A.F. by using the leading fighter of its day, the North American F-86A Sabre, for an attempt on the record. This passed off uneventfully, still at Muroc, with Major Richard L. Johnson, U.S.A.F., flying a normally-equipped Sabre jet to a new record of 670.981 mph on 15 September 1948.

It was at this time that a minor interruption to the development of the speed record occurred, with the start of the Korean War, although the experience of the war showed that the Soviet Union was still lagging far behind in aircraft development. This was in spite of the wholesale use of captured German technology at the end of World War II and the purchase of British engines soon afterwards. The performance of the Mikoyan-Gurevich MiG-15 fighter, particularly when in the hands of North Korean pilots, was such that even propeller types, such as the Hawker Sea Fury, were a match for the Russian aircraft.

Nevertheless, continual development of the Sabre also gave it the first record to be established after the Korean War ended, a little more than four years after its previous record. On 19 November 1952 a North American F-86D Sabre, again without modification, took-off from Thermal, California, and in front of representatives of the National Aeronautical Association and of the F.A.I., flew over a measured course in the Salton Sea at an average speed of 698.50 mph. The Salton Sea was a dry lake some 235 feet below sea level, and one result of this uncannily low altitude was that the speed of sound was approximately 775 mph.

Many British commentators seemed to resent this new record, ostensibly because the performance of the F-86D was no better than that of the Hawker Hunter, Supermarine Swift and de Havilland DH.110, which had been treating Farnborough Air Show crowds to demonstrations of trans-sonic flight. However, the British aircraft were just not ready for a record attempt. Even the Hunter had development problems.

In fact, the Americans were able to prove convincingly that the F-86D's performance was no flash in the pan and that its possibilities were far from being exhausted. Again flying from Thermal, on 16 July 1953, another U.S.A.F. officer, Colonel W. F. Barnes, flew an F-86D to yet another new record of 715.697 mph. However, on this occasion the General Electric J47-GE-17 engine was fitted with an afterburner, which was used for the record flight.

An advantage of flying low as speeds approached Mach 1.0, the speed of sound, was that it was possible to provide the fast runs required for speed records without worrying unduly about the Mach number of the aircraft's aerodynamics. This advantage faded as aerodynamic design improved, and aircraft were then prevented from realizing their true potential because of the effects of air pressure on their performance. The ride for the pilot also became progressively harsher as speeds increased, and altitudes were restricted by the F.A.I.'s maximum of 328 feet (100 metres). The altitude restriction, imposed to assist in verifying performance during the early days of the speed record, had helped the propeller aircraft which lost performance as altitude increased, but it hindered the turbojet.

The Mach number of the F-86D Sabre's first record had been just 0.9, but at altitude a comparable speed would have been in excess of Mach 1.0.

Now came the time for the British aircraft industry to implement its promises, first with the Hunter and then with the Swift. The latter was actually to become the Royal Air Force's first British-built swept-wing fighter aircraft.

A specially prepared version of the Hunter was used for the record attempt. It had a nose that was more pointed than that for the production models, an Avon turbojet and reheat. Later versions of the production Hunter were also to use the Rolls-Royce Avon turbojet, but no operational versions were ever fitted with reheat, and in fact it was found to be desirable to slow up the design somewhat.

It was in this aircraft, painted bright red, that Squadron Leader Neville Duke, a former R.A.F. officer and Hawker's test pilot, took off from the Royal Naval Air Station at Ford, Sussex, on 7 September 1953, for the required four runs over the measured three-kilometre course between Rustington and Kingston Gorse. Little more than twenty minutes after take-off he had pushed the record to 727.6 mph. An unofficial record of 741.66 mph was also established by Duke in this aircraft, and soon afterwards, flying from the Hawker airfield at Dunsfold, Surrey, he flew a 100-kilometre closed circuit triangular course at a record speed for the distance of 709.2 mph.

The news of this record coincided with the news that a team from Vickers-Supermarine were investigating the possibilities of the Castel Benito area in Libya for a record attempt by the Supermarine Swift, which would need to fly at 735 mph to exceed the Hunter's record by the required one per cent. Then, on 21 September, Supermarine's test pilot, Lieutenant Commander Michael Lithgow, R.N. (rtd.), flew a Supermarine Swift F.4 to Libya. He almost immediately started trial flights over a measured three-kilometre course marked by a straight stretch of desert highway between Castel Idris and Azizio on the Azizio Plain.

At last the British had found their 'ideal' site for a record attempt under the optimum conditions, or so they thought at first! The conditions were in fact so favourable that on one occasion Lithgow was nearly boiled alive in his own sweat, with the aircraft's cockpit the saucepan. This was on 23 September, when the cockpit temperature reached an astonishing and hellish 180° F. and Lithgow's cooling suit refused to function properly.

The decision to press ahead with an attempt on the record came the next day, Friday, 24 September, in an ambient temperature of 102° F. It was an exciting flight, not least for the observers from the Royal Aero Club and the F.A.I. on the ground. First, the fuel gauge failed, and Lithgow, uncertain as to the state of his fuel supply, decided to apply reheat when only five miles from the start of the three-kilometre course, instead of ten as originally intended. Then, during one of the four runs over the course, his oxygen mask failed as a result of the excessive sweating of the previous day. As he ripped off the useless piece of facegear while flying at between 70 and 100 feet at well over 700 mph, Lithgow caused the aircraft to jink, startling the observers.

In spite of these misfortunes, however, Lithgow attained 743.6 mph on his first run, followed by 729.5 mph, 745.3 mph, and 730.7 mph (making an average of 737.3 mph) which the F.A.I. homologated as a new record. With the speed of sound at an estimated 796 mph in the prevailing conditions, the Mach number of the Swift's record was 0.926.

Such a performance naturally gave cause for speculation that Hawker's might take the Hunter to Muroc, but it was not to be. In spite of the successful record attempts by the two British aircraft, the North American YF-100 Super Sabre had appeared and was undergoing trials for the United States Air Force, so it was only a matter of time before this aircraft, known to be capable of supersonic speed in level flight, would be used for a record attempt. Most experts at the time believed such an attempt would follow an amendment to the F.A.I.'s archaic altitude restrictions.

In the event, the Swift's record was of short duration, but the aircraft which broke the record was not the Super Sabre. It was time for the United States Navy to assert itself once again, using the prototype of the new Douglas F4D-1 Skyray. Trials with Lieutenant Commander James B. Verdin, U.S.N. as pilot, started over the Salton Sea three-kilometre course on 25 September, just a day after the Swift had broken the Hunter's record. A few days later, a record attempt in the Skyray had to be abandoned due to a faulty fuel gauge; on another day the aircraft produced an average speed of just 742.7 mph, which was 1.6 mph below the required one per cent increase over the previous record. It was therefore not eligible for consideration as a new official record.

After these disappointments the aircraft and Verdin went on to set a record of 753.4 mph (Mach 0.964) on 3 October 1953 over the Salton Sea course in front of representatives from the F.A.I. and the American National Aeronautical Association. The record was the average of four runs at 746.075 mph, 761.414 mph, 746.503 mph and 759.499 mph. The craft was flying at 100–200 feet above the lake level, where the speed of sound was estimated to be 792 mph.

Not surprisingly, after his earlier problems, Verdin was jubilant. 'The plane behaved perfectly,' he told a news conference. 'Turbulence didn't bother me a bit. I feel very fortunate to have been able to make the run.'

He was quite lucky to have had such a smooth twenty-minute flight at such high speed and so low an altitude; this shows the excellent conditions which the Americans enjoyed in California. Verdin's flight was the last official record to be set at low altitude. A record attempt by the YF-100A Super Sabre on 29 October failed to boost the record beyond the required one per cent increase. Lieutenant Colonel Frank K. Everest, U.S.A.F., reached only 755.154 mph.

These performances were enough to depress the pessimists, however, who could not see any hope of a challenge to America's lead appearing in the foreseeable future. The British were, as usual, happily writing themselves off, deciding that by the time a suitable British challenge could be mounted the Americans would have raised the record still further.

As many had predicted, the revised F.A.I. requirements for official speed records were used for the North American F-100C Super Sabres' attempt on the record, which did not come until 20 August 1955. On that day an F-100C took-off from the Edwards Air Force Base in California (with Colonel Horace A. Hanes, U.S.A.F., at the controls) for the flight over the revised 15–25-kilometre course in the Mojave Desert. Flying at 40,000 feet, Hanes made the first run at 870.627 mph (Mach 1.32) and the second at 773.644 (Mach 1.172). The average of 822.135 mph was the first supersonic speed record.

Oddly enough, the Americans at first released only the vaguest of details about the flight, keeping the full story back for the official announcement at the National Aircraft Show at Philadelphia during the weekend of 3–5 September.

In spite of the forecasts of many so-called, and often self-styled, 'experts', the next speed record was indeed to pass to the United Kingdom, which by the early spring of 1956 had also managed to obtain an altitude record, 65,890 feet, using a Bristol Siddeley Olympus-powered English Electric Canberra piloted by Wally Gibbs.

The aircraft for Britain's first and only high altitude speed record was a delta-wing research aircraft, the Fairey F.D.2, more usually and appropriately named the Fairey Delta 2. Powered by a single reheated Rolls-Royce Avon turbojet, the Delta had the unusual feature of a droop snoot, or nose, to aid pilot visibility during landing. In later years it was modified for test flying in connection with the Concorde airliner programme and became the B.A.C. 221.

The course for the record attempt was a measured 15 kilometres between Chichester and Ford, Sussex. R.A.F. Fighter Command radar was used to keep the aircraft on course during the runs over the course and during the final stages of the run-up to the course. Promising trials on 8 March 1956 and on the following day, with a last minute 'dress rehearsal' at 8 a.m. on the morning of 10 March, encouraged all concerned, including the test pilot, Peter Twiss, a former Fleet Air Arm Lieutenant Commander, to press ahead. Accordingly, the attempt was timed for late morning on Saturday, 10 March.

At 11.21 a.m. Twiss took-off in the Fairey Delta. He flew from the Royal Aircraft Establishment at Boscombe Down, Wiltshire, climbed subsonically to 38,000 feet over the New Forest, then started to accelerate over Fawley, near Southampton, and began his final acceleration over Thorney Island. He streaked over the course on the first run at a remarkable 1,117 mph. Then, after going out to sea over Beachy Head to turn, he flew back with acceleration starting over Rottingdean and final acceleration over Angmering. This time the aircraft managed no less than 1,147 mph, making an average speed of 1,132 mph, or Mach 1.731!

This remarkable achievement was an increase of no less than 310 mph, or 37 per cent, on the previous record. The aircraft had covered 240 miles during the flight and had taken twenty-three minutes from take-off to touchdown.

Scoffers said that the aircraft was purely a research machine, whereas the American record aircraft had all been fighters. This was not strictly true. The Douglas Skystreak had not entered operational service, and the Delta inspired no less an aircraft than the Dassault Mirage, early versions of which used Avon powerplants. The record was homologated without difficulty.

The Delta has, to this day, the distinction of setting Britain's last official world air-speed record. The end of 1957 saw the United States in a position to retake the lead again.

On 12 December 1957 Major Adrian Drew, U.S.A.F., took-off from the Edwards Air Force Base in California to fly over a measured course near Los Angeles in a McDonnell F-101A Voodoo interceptor named 'Fire Wall'. Drew, who was later awarded the American D.F.C. by Major General McCarthy, took the record to 1,207 mph. The Voodoo was powered by two reheated Pratt and Whitney J57-1B turbojets; it later entered service with both the U.S.A.F. and the Royal Canadian Air Force.

Another very fast aircraft at this time was the North American F-107A; believed to have been capable of Mach 2.1, it was cancelled with only six aircraft ordered.

Six months later the United States Air Force boosted the record still further, using Lockheed's famous 'missile with a man in it', the F-104A Starfighter. One of these aircraft, flown by Captain Walter Wayne Irvin, U.S.A.F., from Edwards Air Force Base, flew at 1,404.9 mph on 16 May 1958. The F-104A on another occasion established an altitude record of 91,249 feet.

It was at this time that the Soviet Union became a worthwhile challenger for the speed record, although such achievements in the Soviet Union inevitably tend to mean neglect, rather than benefit, elsewhere; such are the effects of a shackled society. A British challenge was not out of the question. The Bristol (later the British Aircraft Corporation) 188 research aircraft was progressing. This plane was being designed for speeds in the region of Mach 3.0 from its two Gyron Junior reheated turbojets. Probably the only real challenger to the 188 at the time was the North American X-15, an air-launched aircraft and therefore ineligible for an attempt on the speed record, but the will to compete seemed to be fading, not so much amongst the British aircraft industry's leaders, but amongst the politicians responsible for commissioning such projects.

Releasing as few details as possible, the Soviet Union claimed the absolute speed record for a flight by Colonel Georgy Musolov on 31 October 1959 in a Sukhoi E-66 delta-wing aircraft. The record was claimed to be at 1,493 mph over a course at Sidorovo Tyvmenskaya, with the fastest run at 1,566 mph.

The Soviet record was short-lived. Major Joseph W. Rogers, U.S.A.F., flew a Convair F-106A Delta Dart from the Edwards Air Force Base on 15 December 1959 to a record of 1,520.9 mph, the average of two runs at 1,535 mph and 1,505 mph. At about this time, an F-104C Starfighter boosted the altitude record to 103,395.5 feet.

After a long interval the United States Navy had found just the aircraft it needed to snatch the speed record from the Air Force. This was the brilliantly successful McDonnell-Douglas F4H-1F Phantom II, designed as a carrier-borne interceptor and fighter-bomber. Lieutenant Colonel Robert B. Robinson, U.S.M.C., flew the Phantom II, named 'Sky Burner', from the Edwards Air Force Base on 22 November 1961 to a new record of 1,606.3 mph. Perhaps not surprisingly, the U.S.A.F. became a Phantom operator in due course. The Air Force used large numbers during the Vietnam War. The aircraft is still amongst the leading combat aircraft. Its export is carefully controlled.

The Soviet Union raised the record slightly on 7 July 1962 when Colonel Musolov flew a Sukhoi E-166 over the Sidorovo course at 1,665.89 mph.

But the record was placed far beyond the reach of the Soviet aircraft on 1 May 1965 by the experimental Lockheed YF-12A, flying from the Edwards Air Force Base of course! The following day, President Johnson was able to announce that the record had been raised by more than 400 mph to 2,070.10 mph by the YF-12A in the hands of Colonel Robert L. Stephens, U.S.A.F. This was achieved at a height of 80,000 feet. 500-kilometre and 1,000-kilometre closed circuit records were also established at the same time with speeds of 1,642 mph and 1,688 mph respectively. These records, by the YF-12A with its two Pratt and Whitney J58 engines, have stood since then, and even the Bristol 188 would have found it very difficult indeed to improve upon the absolute record and impossible to improve on the closed circuit records.

In just sixty-two years aircraft speeds rose from a mere 30 mph to a figure almost seventy times as fast, and other aspects of performance improved too. One feature of the YF-12A, which was designed as a prototype interceptor, and its close relation, the SR-71 reconnaissance aircraft, was speed with range. Few of mankind's inventions have developed so quickly, so safely, and in such a short time as the aeroplane; but there is another side to the story of the speed record during the post-war period, because the fastest aircraft did not set official speed records.

Opposite: The static aircraft park at the 1946 Society of British Aircraft Constructors' display at Radlett. Nearest to the camera is the de Havilland D.H.108, which was the first British aircraft through the sound barrier, and behind that can clearly be seen a de Havilland Vampire, two Hornets and a Dove light transport, followed by a Meteor in R.A.F. camouflage, and a little farther along is a Vickers Valletta, military development of the Viking. In the next row is a Bristol Freighter, which type did much to develop vehicle air-ferry services, a Lancaster bomber converted for turbojet research flying, a Handley Page Hastings military transport and an Avro Tudor airliner. (Photo: Flight International.)

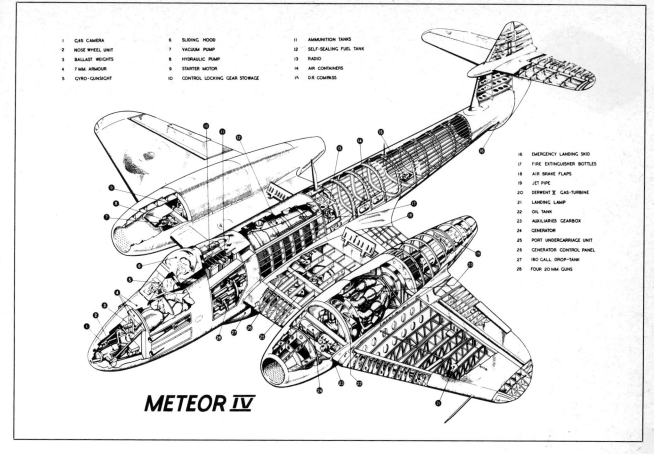

1	G.45 CAMERA	6	SLIDING HOOD	11	AMMUNITION TANKS	
2	NOSE WHEEL UNIT	7	VACUUM PUMP	12	SELF-SEALING FUEL TANK	
3	BALLAST WEIGHTS	8	HYDRAULIC PUMP	13	RADIO	
4	7 MM. ARMOUR	9	STARTER MOTOR	14	AIR CONTAINERS	
5	GYRO-GUNSIGHT	10	CONTROL LOCKING GEAR STOWAGE	15	D.R. COMPASS	

16	EMERGENCY LANDING SKID
17	FIRE EXTINGUISHER BOTTLES
18	AIR BRAKE FLAPS
19	JET PIPE
20	DERWENT V GAS-TURBINE
21	LANDING LAMP
22	OIL TANK
23	AUXILIARIES GEARBOX
24	GENERATOR
25	PORT UNDERCARRIAGE UNIT
26	GENERATOR CONTROL PANEL
27	180 GALL. DROP-TANK
28	FOUR 20 MM. GUNS

METEOR IV

A cut-away drawing of the Gloster Meteor IV, a modified version of which established the first post-war speed records in 1945 and 1946. The earlier versions of the Meteor had the distinction of being the only Allied jet fighter to see combat in World War II. (Photo: Science Museum, London.)

The first post-World War II American record-breaker was the Lockheed F-80 Shooting Star, specially-modified from the standard aircraft shown here. (Photo: Lockheed Aircraft.)

Looking very much the part of a research aircraft, this is the United States Navy's record-breaking Douglas D-558 Skystreak, which flew over the Muroc Dry Lake at 640 mph and then, five days later, at 650 mph. A special rocket-powered development did better still. (Photo: McDonnell Douglas.)

The North American F-86 Sabre set both the last record before the Korean War and the first record afterwards, and although different versions of the aircraft were used, in each case a feature of the record was the lack of special modification. This is an F-86D all-weather version of the Sabre, which set a record of 698 mph in November 1952. (Photo: North American Rockwell.)

Roll out at the then Royal Naval Air Station Ford, in Sussex, for the specially prepared Hawker Hunter, in which Squadron Leader Neville Duke set a record of 727.6 mph in September 1953. The red-painted aircraft included amongst the modifications a reheated Rolls-Royce Avon turbojet. (Photo: Hawker Siddeley Aviation.)

Opposite, top: The Hunter's record was short-lived. This is the record-breaking Supermarine Swift, photographed while in flight over a rather more lush countryside than that of the Libyan Desert, over which it raised the record to 737 mph on 24 September 1953, just seventeen days after the Hunter had set its record. The Swift also used a reheated Rolls-Royce Avon turbojet. (Photo: British Aircraft Corporation.)

Opposite, below: In fact, the Swift's record was even shorter-lived, and before the aircraft returned to the United Kingdom the United States Navy had started pre-record attempt trials with this Douglas F-4D1 Skyray. The Skyray marked the start of a period of ascendency for the delta-wing, punctuated by the success of the North American YF-100 and F-100C Super Sabres and the Lockheed F-104A Starfighter. (Photo: McDonnell Douglas.)

Right, top: The supersonic era for military aircraft was firmly established by the North American F-100 Super Sabre series while in U.S.A.F. and R.D.A.F. service, but before this a YF-100A prototype pushed the speed record to 755.14 mph while flying over the Salton Sea in California. This is a YF-100A on the ground taxying after landing, trailing a very necessary braking parachute. (Official U.S. Air Force Photo.)

Right, middle: A Lockheed F-104A Starfighter, of the type which boosted the record to 1,404 mph in May 1958, although the record aircraft would have flown 'clean'. Successive versions of the Starfighter have been produced during the intervening period, and the aircraft is still in widespread service, mainly in the F-104G development, and F-104As serve with Air National Guard Units. (Photo: Lockheed Aircraft.)

Right: Back to the Delta, or the Delta Dart in this case. This U.S.A.F. Convair F-106A Delta Dart raised the record to 1,520 mph on 15 December 1959, beating an earlier record that year by a Soviet delta-wing aircraft, the Sukhoi E-66. The Delta Dart was a development of the Convair F-102A Delta Dagger and was initially designated the F-102B. (Photo: General Dynamics.)

A lot of hope was placed in the possibility of a record attempt by the Bristol T.188 research aircraft. It was built of stainless steel and at the time of its launch during the early 1960s it was far ahead of any other conventional take-off aeroplane. But nothing happened. (Photo: British Aircraft Corporation.)

Record-breaker meets its admirers. The Soviet record-breaking aircraft have never been seen in the West, and details of the record attempts are usually minimal. However, the Sukhoi E-166 shown here, obviously at an air show, has the distinction of having established an official record of 1,665 mph at Sidorovo in July 1962. Some relationship with or at least similarity to the MiG-21 seems to be evident. (Photo: TASS.)

The 2,070 mph official record holder. Certainly still one of the fastest aircraft taking-off under its own power, although the F-15 is also a Mach 3.0 design. The Lockheed YF-12A was designed to be a formidable addition to the U.S.A.F.'s fighter element, but it proved to be more of a research aircraft. A feature of the YF-12A and the SR-71 reconnaissance version is the tremendous range for such a fast aeroplane. (Photo: Lockheed Aircraft.)

Record holder over a 100-kilometre closed circuit, at 1,615 mph, is this Soviet E-266, a special version of the Mikoyan-Gurevich MiG-23 'Foxbat'. No attempt on the absolute record has yet been made by this aircraft, and perhaps such is unlikely for fear of confirming suspicions held in the West about this aircraft. (Photo: TASS.)

Chapter 9

CIVIL AVIATION AFTER WORLD WAR II

The long-range airliner – early jets – the turboprop – return
of the jet – the supersonic era.

Although many promising civil designs had fallen by the wayside, World War II had given far more to air transport than it had taken. The benefits of wartime aviation were represented not only by technical advance and new aircraft, but also by the pioneering of North Atlantic landplane services through wartime ferry and V.I.P. flights, which brought the realization that air travel for rapid movement of passengers and goods could often be more of a necessity than a luxury. Such developments provided the customer pressure for adequate long-range aircraft from the manufacturers, while to the airlines came encouragement from those who had come to accept air transport as normal during their wartime service.

Large numbers of war surplus aircraft enabled still more ex-service pilots and mechanics to follow their fathers after World War I in starting new airlines and new air services. A terrific momentum was building up on which forward-looking manufacturers, such as Boeing, Douglas, de Havilland and Vickers, could look confidently to the future.

Four types of aircraft were available to civil operators during the post-war period, illustrating the British and the American approach to civil aviation and the vast gap between them.

Britain could virtually offer only converted bombers immediately after the war: the Avro Lancastrian was a converted Lancaster bomber, and the Halton was a converted Handley Page Halifax. There were conversions of the Lincoln and Wellington, and the American Flying Fortress and Liberator. In terms of comfort, the narrow fuselages, lack of pressurized accommodation and adequate sound-proofing made these aircraft a poor substitute for the purpose-built airliner. However, their cruising speeds of more than 250 mph compared well with those of the pre-war period, although the economics of the converted bombers were poor. The United Kingdom–Australia flight took just 63 hours for some 12,000 miles. One of the fastest and the longest flight in the world, it operated twice weekly.

Any parallel between the use of converted bombers after World War II and after World War I must be viewed in the light of the other two aircraft types existing after World War II. The Americans could offer the best of their pre-war aircraft, such as the Douglas DC-3 and the Beech 18, in the knowledge that these were far from obsolete and that large numbers of war surplus aircraft were available for conversion to civilian duties. But the real bonus for the Americans lay in the aircraft developed during the war, the large twin-engined Curtiss C-46, and the new four-engined, long-range Douglas DC-4 and Lockheed 049 Constellation.

The last aircraft type consisted of new civil developments of the Short flying-boats, of which the Solent was the first post-war aircraft and represented

The most successful long-range airliner ever, in prototype form. The Boeing 707 is still in demand sixteen years after its entry into airline service, with low volume production by Boeing, and high prices being paid for used aircraft. (Photo: Boeing Aircraft.)

a determined attempt by the British industry to continue its pre-war effort. Nevertheless, realizing that the trend was to the long-range landplane, Britain was also able to produce civil versions of its Avro York military transport – although this cramped and noisy aircraft showed strong signs of bomber ancestry – and a new twin-engined airliner, the Vickers Viking, built on Wellington bomber aerodynamic surfaces.

Cruising speeds for civil airliners were in the region of 200–250 mph. The Constellation and the DC-4 were at the upper end of the scale, and the DC-3 and Viking operated lower down. A number of pre-war aircraft remained in widespread service, notably the Junkers Ju.52/3M, and these were slower still. In fact, the pressurized Constellation could reach and maintain speeds of nearly 300 mph, but at the cost of some reduction in the range for which this aircraft, like the DC-4, was valued. Passenger comfort on the best aircraft was to pre-war standard, although there were still few pressurized aircraft. Many have since commented on the lack of pressurized accommodation on the DC-4, but it must be remembered that most DC-4s were built as C-54 wartime military transports, for which pressurization would have meant unnecessary expense and possible danger in the event of the fuselage skin being damaged in the rough and tumble of wartime operation.

The DC-4's 250 mph cruising speed from its four 1,200-h.p. Pratt and Whitney Hornet radial engines, its ability to accommodate up to sixty passengers and its 1,800 mile range compared well with the Solent's 200 mph and forty passengers. The Solent's range was the same but at the cost of using four 1,690-h.p. Bristol radials. The flying-boat was still slower than its landplane counterpart and hopelessly uneconomic in the post-war world. It was the DC-4 which introduced so many of the new trans-Atlantic air services, albeit with refuelling stops at Shannon in Eire, and Gander in Newfoundland.

British airlines were virtually forced to buy Douglas DC-3s and a Canadian-built version of the DC-4, the DC-4M or Argonaut, equipped with four 1,760-h.p. Rolls-Royce Merlin engines in a Canadair airframe.

Boeing's contribution to air transport at this time brought a significant increase in cruising speeds. This was the Boeing 377 Stratocruiser of the late 1940s, used mainly by Pan American, T.W.A., B.O.A.C. and United Airlines. Based on the Boeing B-29 Superfortress heavy bomber, the Stratocruiser had the appearance of a landplane conversion of a flying-boat, an impression enhanced by the luxury of the double-deck interior. Its 310 mph cruising speed earns it a place in any summary of civil air speed development. It also enabled it to equal the Constellation's $19\frac{3}{4}$-hour London–New York time while flying via Prestwick against the Constellation's direct route via Shannon.

A new generation of aircraft was making its appearance in the United States for short and medium-haul operations, beginning with the twin-engined Martin 202, and its pressurized successor, the 404, which were soon joined by the Convair 240, a pressurized airliner with a cruising speed of 280 mph from two 2,400-h.p. Pratt and Whitney Double Wasp radial engines. These aircraft marked 1947 as a vintage year in air transport, and they were soon joined by developments of existing aircraft. These included the Lockheed 749 Constellation, with a longer fuselage and better performance than the earlier 049, and the Douglas DC-6 – a four Double Wasp-powered stretched and pressurized development of the DC-4, with a cruising speed of 280 mph.

European aircraft manufacturers were still recovering from World War II, and the few French and Italian civil airliners of the period merely delayed by a short interval the advent of American aircraft in the national air fleets. Even the Dutch found themselves operating first the DC-3 and DC-4, and then the dashing Cv.240 and DC-6, which carried the slogan, 'The Flying Dutchman' on the fuselage. However, it was a K.L.M. DC-6A which won its handicap section's first prize in the 1953 London to Melbourne air race.

The British aircraft industry at this time was frantically building a series of successful military aircraft, but plans were afoot for a number of aircraft capable of countering the seemingly unstoppable American civil challenge. Six new British airliner projects were far advanced during the late 1940s: the Avro Tudor and Handley Page Hermes, both intended to counter the DC-6 and having four engines, the elegant Airspeed Ambassador, a high-wing twin-engined airliner for short and medium ranges, the Vickers Viscount, a four-engined medium-range airliner using turboprop engines, the pure jet de Havilland D.H.106 Comet, and the giant eight-engined Bristol Brabazon. These aircraft, and others, were the result of the recommendations of the Brabazon Committee, which had examined those civil projects most likely to succeed if built by the British industry.

In terms of speed, it was the Comet, which first flew in prototype form at Hatfield in July 1949, which was the most significant aircraft. The Hermes was reliable but unremarkable. The Ambassador, known popularly by its B.E.A. class name of Elizabethan, was merely a suave piston-engined airliner which could have been noteworthy had it entered production with turboprop engines. It did in fact fly as a test bed for every civil British turboprop engine. The Tudor, the only airliner to fly in piston-, turboprop- and turbojet-engined versions, was a disaster in the most literal and painful sense, dragging British South American Airways down with it. A large eight-engined airliner intended to carry upwards of two hundred passengers across the North Atlantic, the Brabazon first flew in prototype form in September 1949, but in spite of the development of a turboprop version it was considered to be too large for the

J

market, and therefore cancelled. A similar fate awaited the even larger Saunders-Roe Princess flying-boat.

The Comet's four 5,000 lb. thrust de Havilland Ghost turbojet engines put the aircraft far ahead of any other airliner, including the de Havilland Mosquito operated on B.O.A.C.'s notorious wartime service to Sweden! A cruising speed of 490 mph and a range of 2,000 miles had to be offset against a passenger capacity of only thirty-six to forty-four. But the comfort of flying in a turbine smooth pressurized cabin at more than 30,000 feet, well above the weather, and the higher productivity of the new aircraft due to its speed, were definite points in its favour.

Originally intended as a three-engined tailless mailplane for the North Atlantic route, with a passenger capacity of no more than six, the Comet had been redesigned as an airliner for the African network of B.O.A.C. as late as 1947. The new aircraft entered service with a flight from London to Johannesburg on 2 May 1952. Canadian Pacific and a French independent airline, U.A.T., also put the aircraft into service; South African Airways operated some of the B.O.A.C. aircraft; and Japan Air Lines and America's Eastern Air Lines had the aircraft on order. A Comet 2 with Rolls-Royce Avon turbojets was being developed as a larger aircraft. Then a series of disasters in 1954 grounded the Comet.

Confidence is everything in civil aviation, and airlines lost confidence in the Comet, even though the cause of the accidents suffered by the aircraft was traced to pressure cracks around the rectangular windows, and this was quickly remedied by fitting round windows with greater strength in their frames. However, the Comet's record cruising speed for civil airliners was far from transient, and Comet 2s entered R.A.F. Transport Command service in 1954 and served reliably until retirement during the late 1960s.

Of all the British aircraft intended for service with the airlines of the 1950s, the one which was an undisputed commercial success was the turboprop Vickers Viscount, the best-selling British airliner ever and the first successful turboprop airliner. The Viscount was extensively developed while in production from the early 1950s until 1964, a period in which 444 aircraft were built. Its four Rolls-Royce Dart turboprop engines gave a maximum cruising speed of 330–360 mph, depending on version, and enabled the aircraft to carry up to eighty-four passengers. Passenger comfort incorporated the smoothness of turbine engines and included large oval windows affording an excellent view of the ground below.

The Viscount was one of the fastest aircraft in airline use during the early and mid-1950s.

Never able to stand still, the American aircraft industry introduced further developments of its Cv.240, the Convair 340 and 440, which represented a slight increase in performance. These long-lasting aircraft were eventually re-engined with turboprops. Those with Rolls-Royce Darts were re-designated the Cv.640. Developments of the Constellation and the DC-6 also appeared, using the new supercharged turbocompound piston engines which marked the ultimate in commercial piston-engined development. The new Douglas was the DC-7, which had a further fuselage stretch, increased range, particularly on the DC-7C version which could fly for more than 5,000 miles without refuelling, and a cruising speed of 360 mph from four 3,400-h.p. Wright Turbocompound engines.

However, the jet airliner had not been abandoned. Boeing's Model 367-80, sometimes known as the 'Dash 80', was developed from the B-47 Stratojet bomber. It first flew in 1954. During the mid-1950s, the 367-80 entered service with the U.S.A.F. as the C-135 transport and KC-135 tanker. It was at this time that the Soviet aircraft industry began to produce civil aircraft worthy of some attention, although in many features they were still lagging far behind the West. The Tupolev Tu-104 twin-jet airliner entered service with Aeroflot in 1956 with a maximum cruising speed of 490 mph – as fast as the Comet 1, but slightly slower than the C-135.

During the 1950s a pattern emerged of piston-engined aircraft capable of cruising at around 360 mph and of turbojet airliners with a cruising speed of about 500 mph. While the early turboprop aircraft had speeds matching those of the piston-engined airliner of the period, these were soon increased by a new generation of turboprop aircraft which, in their range and size, were more directly comparable with the fastest piston-engined types. At the same time, the first of many twin-engined turboprop aircraft also appeared for what has become known as the 'DC-3 replacement market'.

The first of the large turboprop airliners was the Bristol 175 Britannia, which first flew on 16 August 1952 in its medium-range Series 102 version. It used four Bristol Siddeley Proteus turboprops which were so quiet that the Americans soon nicknamed the aircraft 'The Whispering Giant'. Teething troubles delayed the entry of the Britannia into airline service with B.O.A.C., and by this time the first of a larger development, the Series 300, had flown (on 31 June 1956) with up-rated Proteus engines of 4,445 shp each at a maximum cruising speed of 402 mph.

A larger version of the Britannia was later built under licence by Canadair as the CL-44; it used Rolls-Royce Tyne turboprops and had a maximum cruising speed of about 420 mph. A similar performance came from the Russian Tupolev Tu-114 turboprop airliner, introduced to Aeroflot service in 1960 for very long distance services requiring its 5,500 mile range. The two-hundred passenger

Tu-114 used four 14,705-shp Kuznetsov NK-12MV turboprops for a maximum cruising speed of 420 mph.

At a time when the aeroplane was no novelty, it is interesting to see that many of these aircraft ran into difficulties. The Comet 1 had been a disaster, the Boeing 367-80 developed many minor faults at the start of flight testing, and the Britannia at first had the unwelcome tag of 'around the world in eighty delays'! All was well in the end, and these three particular aircraft became known as reliable and dependable machines, although this assessment was reserved for only the later version of the Comet. The Russians kept their mistakes to themselves and were indeed reticent about even natural disasters.

Such high speeds as those of the Britannia and Tu-114 were not matched by the 'DC-3 replacement' turboprop airliners, of which the first was the Fokker F-27 Friendship, which first flew in prototype form in November 1955. This aircraft and others in the same category, such as the Handley Page Herald, which first flew in March 1958, the Hawker Siddeley, or Avro, 748, which first flew in June 1960, and the Japanese Nihon Aircraft Manufacturing Company YS-11, all had cruising speeds in the 275–300 mph range. All used two Rolls-Royce Dart turboprops.

Military transport turboprop aircraft, including the twin-boom Armstrong-Whitworth Argosy, the Lockheed C-130 Hercules, and the Douglas C-131, also fell into the non-record bracket. The fastest, the C-131, cruised at less than 400 mph.

A range of medium-haul turboprop airliners, larger than the Viscount, also entered airline service during the late 1950s, including the Vickers Vanguard, the Lockheed L-188 Electra and the Ilyushin Il-18. All three aircraft possessed a range of 1,800–2,400 miles thus qualifying them as medium-range, although ten years earlier this would have been considered to be definitely long-range. The fastest of these aircraft, and possibly the fastest turboprop airliner, was the Vickers Vanguard, equipped with four 5,545-shp Rolls-Royce Tyne turbo-props giving a cruising speed of 430 mph, a range of up to 1,830 miles and a capacity of up to 139 passengers. The Vanguard first flew on 20 January 1959.

Having devoted so much time and effort to the development of propeller aircraft with fairly high cruising speeds, passenger accommodation and comfort together with a good range, the manufacturers of all but the short-range airliners were in for a major disappointment. What might almost be described as a 'second jet age' dawned in 1958 with the advent of the de Havilland Comet 4 and Sud Aviation Caravelle. The Comet 4 appeared in long- (the 4 and 4C) and medium-range (4B) versions; the latter was the faster aircraft with a maximum cruising speed of 526 mph from its four 10,500-lb. thrust Rolls-Royce Avon 525 turbojets.

Sud's twin-jet Caravelle, with the then new feature of tail-mounted engines, was the first and only French post-war airliner to become a commercial success. The aircraft first flew in prototype form on 27 May 1955. It entered regular airline service in 1959. A variety of Rolls-Royce Avon and Pratt and Whitney JT8D powerplants were used, giving the Caravelle a cruising speed of 490 mph and the Super Caravelle of 1964 a cruising speed of 518 mph. This aircraft could justifiably be described as the first purpose-built short-haul jet, although its performance over short stage lengths, such as London to Paris, was such that the turboprop Vanguard normally managed a better journey time than the Caravelle's early versions!

The Comet was obsolescent by this time, however. It was very much an aircraft of the early 1950s in concept. In spite of some remarkable orders from airlines in Africa, the Middle East and Latin America, who doubtless appreciated its ruggedness (although one airline managed to write off half its fleet of six aircraft very quickly), the Comet was almost completely overshadowed by the faster Boeing 707, Douglas DC-8 and Convair Coronado.

Of these three aircraft, the Boeing 707 was the first to appear. It was a development of the original Model 367-80 of 1954 and the military C-135 and KC-135. Few aircraft can have been built in as many differing versions as the 707 and its shorter range close relative, the 720, with differing fuselage lengths, passenger, cargo, or quick-change passenger-cargo versions, and Rolls-Royce or Pratt and Whitney turbojets. Later versions had Pratt and Whitney turbofans.

A cruising speed of 580 mph from the 707 and the DC-8 – available in fewer versions than the Boeing, but with a remarkably similar performance and appearance, and much the same choice of powerplants – was a remarkable stride forward. For the passenger it could mean London–New York in under seven hours, but the range of these aircraft extended up to 6,000 miles, and up to 190 passengers could be carried depending on version and interior layout. Four Rolls-Royce Conway turbojets of 17,500-lb. thrust, Pratt and Whitney JT4 turbojets of similar output, or, later, Pratt and Whitney JT3D turbofans of 18,000-lb. thrust, mounted in under-wing pods, provided this creditable performance.

An extensive stretch of the Douglas DC-8 in 1966, introducing the DC-8-60 or 'Super Sixty' series, with accommodation for up to 250 passengers or an 8,000-mile range, or, on certain versions, both, did not bring any further improvement in cruising speed.

Convair's Coronado was a faster aeroplane than either the 707 or the DC-8, although it had a smaller payload and a shorter range. The first version of the Coronado, the Cv. 880, first flew in January 1959 with four 11,200 lb. thrust General Electric CJ-805-3B turbojets; the developed 990A of 1960 used

16,090-lb. thrust turbofans for a maximum cruising speed of 625 mph. The small size of the aircraft, with a maximum capacity of 106 passengers, prevented its collecting anything like the number of orders won by the 707 or DC-8. The larger aircraft were necessary for the vast boom in air travel which followed the introduction of jet aircraft, even though many had predicted that the new airliners would be too big for the market.

Unlike the development of the modern piston-engined airliner before the war, when the initial progress was made at the lower end of the scale for size and range, and developments then moved upwards to larger and longer range aircraft, the reverse took place with the turbojet. The Vanguard had managed to achieve faster London to Paris times than the early Caravelle because the latter aircraft had to waste time climbing to an economic operating height. Improvements in airframe and engine design soon resulted in improved climb performance, however, and the development of the turbofan, with the initial set of fan blades larger than the compressor blades and in effect acting as a ducted propeller, soon brought together the advantages of the turboprop and the turbojet, giving the turbine engine a wider range of performance than hitherto.

Although the Boeing 720 was a medium-range airliner, it was a variation on the 707 theme. The first of the modern medium-range airliners came with the trijet Hawker Siddeley Trident and the Boeing 727. The Trident One, which entered service with B.E.A. in March 1964, had a cruising speed of 600 mph, but the Trident Two of 1968, with up-rated Rolls-Royce Spey Mk.512 turbofans of 11,930-lb. thrust each, had a cruising speed of 625 mph and a range of 2,450 miles with a 115-passenger load. The Boeing 727-100, which also entered airline service in 1964, a month earlier than the Trident, has about the same cruising speed as the Trident Two, but with a 131-passenger capacity and 2,000-mile range from three 14,000-lb. thrust Pratt and Whitney JT8D-1 turbofans; the improved 727-200 brought an increase in capacity, but not in speed.

These new aircraft thus became the fastest civil airliners, while the Trident had the additional distinction of making the first fully automatic landings. The 727 has been the most successful airliner of the jet age with more than a thousand aircraft sold.

The appeal of the jet airliner compared with even the turboprop was such that the aircraft industries of the world soon found a willing market for short-haul airliners, as airlines vied with one another for what has been described as 'jet appeal'; doubtless they had in mind those passengers who had become tired of the charms of the stewardesses.

However, the need to be able to operate from relatively short runways has meant that the in-flight performance of the short-haul jets has been less exciting than that of the medium-range aircraft. The first of the new airliners were the British Aircraft Corporation One-Eleven, using two Rolls-Royce Spey turbofans, and the McDonnell Douglas DC-9, using Pratt and Whitney engines. Both aircraft had tailmounted engines like the Trident and Boeing 727, but performance was well down with cruising speeds of around 540 mph. The more successful of these two aircraft was the DC-9, although more than two hundred One-Elevens have been sold. A number of versions of both aircraft have been built, with seating capacity varying from 70 to 120.

A faster short-haul aircraft was the Boeing 737, in effect a mini-707 using two under-wing jet engines, but, like the 727, using as many 707 components as possible. The engines of the 727 and 737 are the same. The Boeing contender for this market could cruise at 580 mph, but it has the air of a larger and longer-range aircraft than the DC-9 or One-Eleven.

Smaller still and slightly slower have been the latest small jet airliners, including the Fokker F-28 Fellowship, basically a sixty-five or seventy-seat successor to the F-27 Friendship, and the small VFW-Fokker 614, which has the unusual feature of engines mounted in wing overpods to minimize the noise heard on the ground at the time of take-off and climb. Slowest of the aircraft in this category has been the Russian Yakovlev Yak-40, which has three tail-mounted turbofans giving a cruising speed of little more than 400 mph.

The fashion for tail-mounted engines held for long enough to allow two long-range airliners conforming to this layout to enter service; these were the British Vickers VC.10, along with its larger development, the Super VC.10, and the remarkably similar Russian Ilyushin Il-62. These aircraft could not match the medium-haul aircraft for speed, or the 707 and DC-8 for range, although the British aircraft provided an excellent take-off and landing performance, particularly from high altitude airfields, and better than average passenger comfort. A cruising speed of 580 mph could be attained by a Super VC.10 from its four Rolls-Royce Conway turbofans of 22,500-lb. thrust each.

Russian airliners do not tend to provide sparkling performances, and the Tupolev Tu-13 and Tu-154 are inferior to the DC-9 and Trident, respectively, in terms of cruising speed. There is no evidence to suggest that the Il-62 is any faster than the British and American competition.

The two most recent developments in air transport have proved on the one hand to be as great as the advent of the modern airliner during the early 1930s, and on the other to be as big a step forward as the introduction of the jet airliner. The Boeing 747, hailed as the 'jumbo' jet because of the tremendous girth of its fuselage, with seating of between eight and ten abreast, has introduced a new concept of spaciousness into airliners which has been copied by the so-called 'airbuses' – the Lockheed L-1011 TriStar, the McDonnell Douglas DC-10 and the Airbus Industrie A.300B. The supersonic airliner, typified by the BAC-Aerospatiale Concorde, has simply meant a doubling of aircraft speeds in civil aviation!

The first flight of the Boeing 747 occurred on 9 February 1969, without a prototype stage. Although normally carrying about 350 passengers, versions of this aircraft, with its four 43,500-lb. thrust Pratt and Whitney JT9D-3A advanced technology turbofans on the early versions, and 47,000-lb. thrust on later aircraft, can carry up to 534 passengers over long ranges. No speed records have been set. The 747 has a cruising speed of around 610 mph and a fairly sluggish climb rate on take-off.

The fastest subsonic civil cruising speeds can be matched by the airbuses, however, which can also carry up to four hundred passengers in the case of the Lockheed TriStar and McDonnell Douglas DC-10. These aircraft also have the eight–ten abreast seating of the 'jumbo', but with three engines, two on under-wing pods and a third in the fin. The TriStar, which is the quietest airliner yet, uses three 42,000-lb. thrust Rolls-Royce RB.211 advanced technology turbofans for a range of some 3,000 miles and a cruising speed of 625 mph. The DC-10 uses either 51,000-lb. thrust General Electric CF6-50C or 50,000-lb. thrust Pratt and Whitney JT9D-25W advanced technology turbofans for a range of up to 5,000 miles.

The advanced technology turbofan uses a double or triple spool to enable the turbines to rotate at the optimum speed for each set, thus giving maximum efficiency at some cost in the traditional basic simplicity of the turbine engine. The twin-engined European airbus, the Airbus Industrie A.300B, uses two under-wing mounted CF6 turbofans and is slightly slower than the American aircraft; it is intended to carry a smaller passenger load over a shorter range.

Apart from their not unremarkable turn of speed, this latest generation of aircraft has resulted in a further improvement in passenger comfort. The wide fuselages are not to everyone's liking, however, although sneers at 'cinema seating' layouts were only heard during the early months of operation. Having provided passengers with high-speed flight which is reliable and comfortable, airlines have now had to provide film shows and piped music for passengers on long distance flights. The remarkable paradox is that the tedium must be far less than that of labouring across the North Atlantic at a droning 300 mph!

Yet, in the past every airliner of a generation has aimed for a speed within a certain bracket, and it is only now that a wide differential between the fastest and the standard long-range airliners has become acceptable. Today, real speed in civil aviation is represented by the supersonic airliner, or SST.

The first airliner to fly faster than the speed of sound was the Tupolev

Tu-144 while in prototype form, but this aircraft, with its four 38,580-lb. thrust Kuznetsov NK-144 turbofans with reheat has run into many and varied difficulties during its development phase. Regular airline service with Aeroflot is planned for 1974, however, with the aircraft carrying between 126 and 140 passengers on trans-Atlantic and trans-Siberian supersonic flight, at a cruising speed of around Mach 2.0 (about 1,300 mph).

The Tu-144's competitor, and the second airliner to exceed the speed of sound, is the collaborative BAC-Aerospatiale Concorde, which has four 38,050-lb. thrust Rolls-Royce/Snecma Olympus 593 turbojets with reheat, accommodation for about 110 passengers and a range of 3,500 miles. Concorde is also intended to cruise at 1,300 mph when it enters airline service with B.O.A.C. in 1975, although on test it has shown a maximum speed of 1,400 mph plus.

Concorde is probably the faster of the two aircraft, and as such is the real record-holder of airliner speeds. It could cut the $6\frac{1}{2}$-hour trans-Atlantic flight time to $3\frac{1}{2}$ hours. Certainly, if such aircraft can succeed, it is likely that success will go to Concorde, which has better operating economics from its twin-spool Olympus turbojet, requiring reheat (hitherto a feature only of military aircraft) during acceleration only. The Tu-144's turbofans on the other hand cannot provide supersonic cruising power on their own and need to be aided by reheat during flight. This means that the Tu-144 has the penalty of maintaining four ramjets, in effect, in the jet efflux pipes.

The contrast in cruising speeds between Concorde and the Airco D.H.4A is considerable – (a thirteen-fold increase), and the difference between the D.H.4A and the Boeing 747 is that of an increase in passenger capacity of a multiple of 250.

Even more considerable would have been the difference between the D.H.4A and the proposed American supersonic airliner, the Boeing 2707. Boeing was selected to develop this Mach 2.5 airliner as the result of a design competition for U.S. Government support between Boeing and Lockheed. Lockheed proposed a fixed-wing aircraft and Boeing a variable-geometry type, with the wings swept back in flight, but with the angle of sweep reduced for low speed flight and for take-off and landing. Passenger capacity was to be of the order of 250, and a new heat-resistant material was to be used in the aircraft's construction.

Problems during development, delaying the aircraft and increasing its cost, led to cancellation of the 2707, and no doubt the Americans, using advanced technology in their aircraft, were dismayed by the heavy cost increases (from below £200 million to £1,000 million) incurred on the 'conventional' Concorde.

However, American interest in supersonic commercial flight is far from dead and can hardly be described as completely dormant. Experience gained from continued studies, and from the North American Rockwell B-1 bomber, will doubtless ultimately be used for supersonic commercial flight. The Americans will not surrender a part of the airliner market without a struggle, and failure to develop a supersonic airliner would be surrender.

In the meantime, a further improvement in subsonic speeds can be expected from American studies of area ruling and supercritical wings, maximizing lift and minimizing drag. These features could give the next generation of long-range airliners cruising speeds of about 660 mph, although elsewhere work on reduced take-off and quiet take-off aircraft will tend to force a stabilizing of airliner speeds.

It must certainly not be forgotten that the few European airliners in production and under development leave much of the airliner market to the American aircraft industry, and it is only in supersonic commercial flight that the Americans lag behind at present. In every other field they lead!

A well loved aircraft with a good turn of speed, the Boeing 377 Stratocruiser brought flying-boat standards of comfort to the airways after World War II, although it was very definitely a landplane. This two-deck aircraft had its origins in a military design, and operated with United, Pan American, T.W.A. and B.O.A.C. It was at its best on services with a high proportion of first-class traffic. (Photo: Boeing Aircraft.)

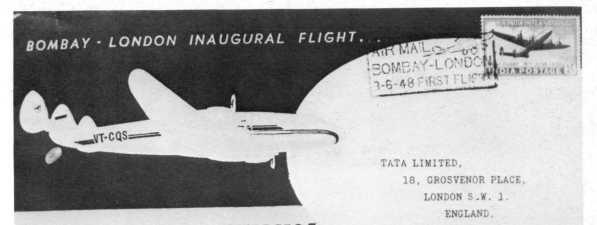

BOMBAY · LONDON INAUGURAL FLIGHT...

AIR·INDIA INTERNATIONAL

TATA LIMITED,
18, GROSVENOR PLACE,
LONDON S.W. 1.
ENGLAND.

A must for the newly-independent nation is a national airline, regardless of cost and need! The first British colony to be awarded independence, if one ignores the 'old white Commonwealth territories', was India in 1948, and by June an Air-India Lockheed Constellation had inaugurated a Bombay–London air service carrying mail and passengers. This is a first flight cover.

Nearer to reality was the de Havilland Comet 1, a graceful achievement for the British aircraft industry which ran into problems leading to its grounding just as it was on the brink of worldwide success. The problems, severe in their results, were not too difficult to rectify, but by that time customer confidence had gone. (Photo: British Airways.)

Repeating its pre-war air race successes, K.L.M. won the handicap section of the October 1953 London to Christchurch Air Races with this aircraft, Douglas DC-6A PH-TGA, 'Dr Ir. M. H. Damme'. The aircraft took 49 hours 57 minutes for the journey, of which flying time was 47 hours 42 minutes for the 13,064 miles. The DC-6A was a pressurized and stretched development of the DC-4. (Photo: K.L.M.- Royal Dutch Airlines.)

Only the Royal Air Force used the de Havilland Comet 2, which incorporated modifications to the windows as well as other improvements of the Comet 1, including the use of Rolls-Royce Avon turbojets. The Comet 2 first flew in 1953 and entered service in 1956. It earned a reputation for reliability and safety during a long service career, and shared with the Tupolev Tu-104 the reputation of being the fastest jet transport for most of the time. (Photo: Hawker Siddeley Aviation.)

A contemporary of the Vickers Vanguard, which it closely resembled in some aspects of its appearance, particularly the tailplane, the Ilyushin Il-18 marked the Russian aircraft industry's entry into the large turboprop airliner field, with speeds of 360 mph and upwards on medium-range services. (Photo: LOT.)

The first trans-Atlantic jet airliner services were operated by B.O.A.C. with de Havilland Comet 4 jet airliners on 4 October 1958. A stretched fuselage, more powerful engines and other minor modifications cannot hide the resemblance to the ill-fated 1; but the Comet 4 soon established a reputation for ruggedness and safety, even though slower than the rival American aircraft. (Photo: British Airways.)

Slightly faster than any of the other aircraft in the first generation of jet airliners during the late 1950s, the Convair Cv.880 Golden Arrow or Coronado shared the Comet's lack of success. This was probably due, in each case, to a narrower fuselage than that of the Boeing 707 and Douglas DC-8. (Photo: General Dynamics.)

One of the features of the Douglas DC-8 was the offer of extensive fuselage and range stretches in the so-called 'Super Sixty' series, which was built during the mid-1960s. These aircraft were no faster than their original and smaller sisters, but in minimising refuelling stops they could reduce journey time. (Photo: McDonnell Douglas.)

No increase in civil speed, but an increase in the airfields and air services which could be used by jet aircraft came during the 1960s with the advent of the 'bus stop' jets, such as the twin-engined B.A.C. One-Eleven and Douglas DC-9. Here a B.A.C. One-Eleven 475, a development built to operate from short gravel runways and 'use any airport the Dakota can', demonstrates its capabilities by taking-off from Tingo Maria in Peru. (Photo: British Aircraft Corporation.)

Many Soviet civil types look like the unhappy offspring of a shotgun marriage between an airliner and a military reconnaissance type, this Tupolev Tu-134 is a case in point. Basically, the Tu-134 is an aircraft in the same short haul category as the One-Eleven and DC-9. (Photo: LOT.)

Top: Much like a Vickers VC-10, in appearance, the Ilyushin Il-62 is now the Soviet Bloc's main long-haul airliner, in spite of a number of mysterious accidents. (Photo: Aviaexport.)

Bottom: Soviet airliner development reflects that of the West, although not always successfully. This is a Tupolev Tu-154 trijet, capable of carrying 150 passengers over ranges of upwards of 2,000 miles. It is frequently described as the 'most exportable Soviet jet so far', but none have in fact been sold outside the Soviet Bloc! (Photo: Aviaexport.)

Opposite: With the distinction of having brought jet travel and 600 mph speeds to many air routes, the Boeing 727 also has the record for turbojet airliner sales, with orders soaring past the 1,000 mark in September, 1972. (Photo: Boeing Aircraft.)

Above: A Tupolev Tu-144 prototype 'on show'. Later prototypes incorporate a retractable fore-plane, sometimes described as a 'canard', to aid low speed handling, which is believed to be poor. Similarity to the Concorde can be seen, although there are few handling difficulties with the latter. In spite of at least one major accident, the Tu-144 is in full production for Aeroflot. (Photo: Aviaexport.)

Top, right: The mock-up of the unsuccessful Boeing 2707 after the decision was taken to abandon variable-geometry. The failure of this project was due to the decision to press too far ahead too quickly, resulting in technical and cost acceleration problems which meant a withdrawal of official support. No doubt, sufficient funds would have meant success. (Photo: Boeing Aircraft.)

Right: One of the Concorde prototypes, in this case a British-built aircraft, flying fairly high but at relatively low speed. The Concorde's Mach 2.0-plus speed is the same as that of many interceptors. Consequently the speed of the fastest airliners will have more than doubled on entry into airline service in 1975. There have been few technical snags in the Concorde programme. (Photo: British Aircraft Corporation.)

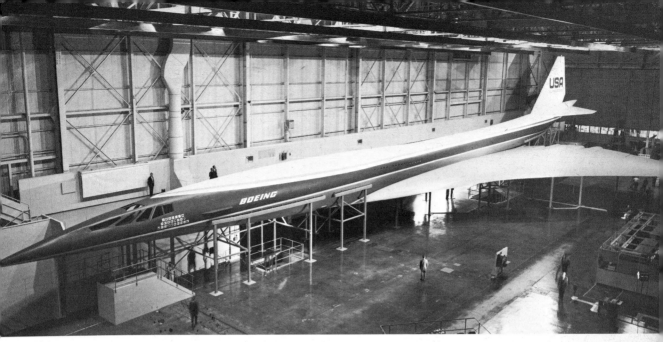

The longer range series 30 development of the McDonnell Douglas DC-10 airbus. The airbus generation of aircraft is introducing high capacity and low noise operation to the world's airlines and airports, while at least maintaining the speeds of the earlier jet airliners and sometimes bringing a marginal increase. (Photo: McDonnell Douglas.)

A new era dawned on air transport with the appearance of the Boeing 747, promptly named the 'jumbo' jet. No real increase in speed resulted, although the aircraft operates the longest non-stop schedule in the world, that of 7,000 miles between London and Johannesburg for South African Airways, on which run it carries 250 passengers in considerable comfort. (Photo: South African Airways.)

Chapter 10

ONLY THE FASTEST

Air-launched aircraft – Bell XS-1 and X-2 – North American
X-15 programme – Mach 6 plus.

One of the ironies of the whole subject of speed in the air, and of the speed record in particular, is that the fastest manned aircraft is not eligible to hold the official world air speed record. This is because the aircraft in question, the North American X-15-A rocket-powered aircraft, which is air-launched from a specially modified Boeing B-52 Stratofortress jet bomber, would have considerable difficulty in finding and then adhering to a measured 15/25-kilometre course. It would also run foul of the requirement that flight should be level within a margin of 100-metres altitude.

Air-launched rocket aircraft are in many ways a hybrid between the conventional aeroplane and the space rocket, with the crew of the latter having little control during take-off through the earth's atmosphere. The X-15-A is the final development of the North American X-15 programme, which in turn was the heir to substantial work by the United States Air Force and the National Aeronautics and Space Administration during the post-World War II period.

The rocket-powered aeroplane was actually an earlier arrival than the turbojet. Both the first two rocket-powered aircraft were German-built and based upon contemporary sailplane design. The first was the tail-first 'Ente' or 'Duck', a name which doubtless referred to the canard configuration of the aircraft. It first flew, for just three-quarters of a mile, on 11 June 1928, near the Wasserkuppe Mountain.

The second rocket-powered aircraft was built and flown by Fritz von Opel on 30 September 1928 at Rebstock, near Frankfurt. It was of more conventional design, but the tailplane was mounted on outriggers.

Neither of these early aircraft posed any challenge to the supremacy of the radial and in-line piston engines of the day. Both were far from being practical aeroplanes.

It was not until the advent of the Messerschmitt Me.163 Komet of World War II that the rocket engine could demonstrate its true potential, aided by the successful application of liquid-fuels. This type of powerplant was in reality not far from the 'light the blue touchpaper and retire' stage, and the Komet, designed as an interceptor to counter heavy Allied bombers, was amongst the first practical and 'modern' rocket-powered designs. A parallel between the Komet and later American designs was that, although usually launched from a ramp, the German aircraft was sometimes towed into position by a bomber aircraft before being released. Operational speeds were in the region of 570–600 mph until on 2 October 1941, after being towed to 13,000 feet, test pilot Heini Dittmar reached 623 mph in level flight.

One of the earliest American rocket-engined types was the Northrop MX-324 research aircraft, which after trials as a glider flew under rocket power for the first time on 5 July 1944.

The fastest manned aircraft so far is the North American X-15, which has attained the astonishing speed of 4,534 mph, although this cannot be ratified as an absolute official speed record. This photograph shows an X-15 streaking across the Rogers Dry Lake in California before beginning the final approach to the Edwards Air Force Base. (Photo: U.S. Air Force.)

Bell Aerosystems-built planes made many of the early flights. The first was the XS-1 (Experimental Supersonic 1), which first flew under power on 8 December 1946, after being carried aloft by a Boeing B-29 Superfortress bomber. A straight-wing aircraft, it had a thin wing for the time and was designed for a speed of 1,700 mph, although this was never reached because of the poor endurance while flying at full power. Nevertheless, in October 1947 it became the first manned aircraft to exceed the speed of sound (Mach 1.0), and on another occasion it reached Mach 1.5. Like most aircraft of its type, the XS-1 was able to glide in to land, shutting-off its four nozzle 6,000-lb. thrust rocket engine if the fuel hadn't been exhausted first!

A second aircraft, the Bell X-1A, reached Mach 2.5 on 16 December 1953, although the United States Navy's Douglas D-558-2 Skyrocket had reached Mach 2.01 even earlier, on 21 November 1953 while being flown by test pilot Scott Crossfield. The Skyrocket was a development of the record-breaking Skystreak, substituting a swept-wing for the straight wing of the conventional aircraft and initially using mixed turbojet and rocket propulsion. A conventional take-off enabled the Skyrocket to reach Mach 1.4 in level flight, but when the turbine was removed and the aircraft was launched from a B-29 Superfortress at 32,000 feet, the unofficial record of Mach 2.01 was possible.

A Mach 3.0 Douglas design of the early 1950s, the X-3, was unsuccessful, but the Bell X-2 raised the rocket aircraft record further. One of the first stainless steel aircraft, the X-2 managed to exceed Mach 3.0 by flying at 2,148 mph and at 126,200 feet on 27 September 1956, a record which was to stand for four years while North America built three X-15 research aircraft.

The much-prized Mach 3.0 was a goal for many aircraft, whether conventional or otherwise, but it was little more than a starting point for the North American X-15, powered by a liquid oxygen and ammonia rocket engine. An exhaustive programme of test flying started early in 1960, when the aircraft was carried to an altitude usually well in excess of 40,000 feet by a specially modified Boeing B-52 Stratofortress bomber flying from the Edwards Air Force Base in California.

The first four flights under the B-52 were made without the release of the X-15, while the aerodynamics of the composite aircraft were investigated, and gliding trials followed before the application of power from the Reaction Motors LR11 powerplant. The LR11 was in fact an interim engine while Reaction Motors worked on the definitive LR99-RM-2 single-chamber throttlable rocket of no less than 57,000-lb. thrust at sea level.

Although a number of pilots were put on to the X-15 programme during its seven year duration, the two principal pilots were Major Robert White, U.S.A.F., and Joseph Walker of N.A.S.A. The team and its three aircraft — four, if the B-52 is counted – were based on the almost inevitable Edwards Air Force Base in California, a primarily U.S.A.F. base which has been the scene of more record flights, whether speed or altitude, than any other airfield in the world.

One of the first really high speed flights of the X-15 took place on 4 August 1960. Joe Walker flew the aircraft from the B-52, which had carried the X-15 from Edwards to an altitude of 45,000 feet. Applying full power after release, Walker put the X-15 into a climb to 78,000 feet before levelling out. After just four-and-a-quarter minutes of full power he reached 2,196 mph (Mach 3.32), beating the Bell X-2's record. He then returned to Edwards and touched-down at 200 mph!

A problem with an aircraft which can fly as fast as the X-15 is the effect of heat on the structure. This is one reason for the use of metals such as stainless steel and other costly alloys, which often have the drawback of being difficult to work. Recording the temperatures experienced is a problem in itself. Early in the X-15 programme the ingenious solution of using heat sensitive paint on the nose of the aircraft was introduced.

The first occasion on which the heat sensitive paint was used, together with some aerodynamic modifications to the nose, was 7 February 1961, when Bob White took the aircraft to 2,275 mph (Mach 3.50).

By this time the more powerful Reaction Motors LR99-RM-2 powerplant had been fitted to the aircraft and was settling down well in service. The way was clear, however, for N.A.S.A. to proceed with its programme of studies for which test flying with the LR11 had really been no more than a warming up exercise, in spite of the new speed records set. One of the earliest flights with the new engine came on 7 March 1961, when the aircraft exceeded the fastest speed attained by the LR11 engine by no less than 605 mph, levelling out after release from a B-52 and climb at 75,000 feet, and using just half power to reach 2,905 mph, or Mach 4.43. Visible evidence of this achievement was on the nose-paint of the X-15 – the 'hot nose' had changed its colour during the flight from green to blue, then to yellow, through black to a final brown, indicating a temperature on the surface of the aircraft of no less than 680° F.

The significance of this half-power performance was that it showed a considerable potential in the X-15 design, and that its design speed of Mach 5.0 was unduly conservative. This was subsequently revised upwards to Mach 6.0, about 4,000 miles an hour, but the aircraft still kept pushing the record upwards. It is only a pity that a conventional take-off could never be attempted which, with radar guidance for the pilot, could have made an official assault on the speed record possible.

1961 was far and away the X-15's best year in terms of speed records and steady progress, but in altitude too the aircraft was setting a pace which no other aircraft has managed to rival, and will not until the advent of the space shuttle. On 21 April Bob White took the aircraft to a record of 3,140 mph using a short burst of full throttle on the LR99 engine. Just a few weeks later, on 25 May, Joe Walker, N.A.S.A.'s chief test pilot, reached 3,370 mph (about Mach 5.0).

One of the more exciting flights in the mostly uneventful X-15 programme came on 28 September, when the aircraft was being flown, not by a N.A.S.A. or U.S.A.F. pilot, but by Commander Forrest Petersen, U.S.N. After release, the aircraft climbed to 82,000 feet, where it reached a speed of 3,545 mph and the exterior temperature soared to 1,100° F. It was so hot that, according to Petersen, smoke from burning paint on interior surfaces came 'right out under the front of the instrument panel into my face'.

Nevertheless, by the end of that month the three aircraft in the test programme had chalked up a total of more than forty flights between them, and, although they didn't release all the details of the programme, N.A.S.A. was able to claim that the aircraft was proving to be a much more useful and adaptable research tool than had been originally envisaged. There were many more inputs for research, both military and civil, than had been hoped for.

Much of this research was connected with manned spacecraft development, including control systems and some work with a bearing on the eventual space shuttle programme, although other aircraft, such as the Northrop Lifting Body, also had contributions to make. Research into reconnaissance techniques and possibilities was also conducted, and the sciences of metallurgy and aerodynamics were other beneficiaries from this programme. Gone were the days of speed and altitude for their own sake, or even as gestures to promote the sale of aircraft.

Another moment of excitement came during a flight by Major White on 9 November 1961. After flying at 4,039 mph (Mach 6.04) at between 90,000 and 100,000 feet, thus setting a new X-15 record, the outer layer of the right hand windscreen crazed, although fortunately the shattered panel remained in its frame. Modifications were made to the windscreen before White attempted to set an altitude record (also unofficial) of 250,000 feet. He actually reached 247,000 feet, although he thought he had reached 250,000 feet before beginning his descent.

The speed of 4,093 mph started speculation about the X-15 having reached its full potential. After some modifications of a relatively minor nature, however, the aircraft, as the X-15A-2, was able to go still further in 1966. Even before this, on a flight designed to test the aerodynamic stability of the X-15 at a 23 degree angle of attack, Walker flew at 4,104 mph (Mach 6.09) on 27 June 1962 while at 96,000 feet.

The first complete outing of the X-15A-2 did not come until after Walker's death. The B-52 launch was still adhered to, and on 18 November 1966 Major William Knight, U.S.A.F., raised the record by a further 55 mph to 4,159 mph.

Almost the last high speed flight of the X-15A-2 came in 1967, about a year after the 4,159 mph record. This flight, on 3 October, was the last of the speed records from the X-15 programme, and like that of the Lockheed YF-12A, it still remains intact. In just an eight minute flight from launch, the aircraft, with Major William Knight as pilot again, reached the staggering speed of 4,534 mph (approximately Mach 6.72) at 99,000 feet, an achievement made possible by adding 13,500 lb. of anhydrous ammonia to the fuel load, giving an extra 60 seconds of engine burn time.

The flight had its minor moments of drama. The release button had to be pushed twice before the aircraft fell away from the giant B-52, and a warning light indicated that the motor was overheating just before the end of the high speed run! The overheating at any rate could hardly be considered surprising.

The programme did not end until 1969, by which time a number of further altitude records had been established. White's record was raised by almost 50 per cent to 354,200 feet (more than 67 miles above the earth's surface). All in all, the three X-15s made just 199 flights between them during their nine-year active life, and the total airborne time, excluding time spent under the belly of a B-52, amounted to just 30 hours, 13 minutes — less than half a week's flying time for a modern jet airliner!

The American rocket-powered aircraft research programme really got underway with the flights of the Bell XS-1, which was air-launched from a Boeing B-29 Superfortress. Although the aircraft was designed for a speed of 1,700 mph, this was never reached. Nevertheless, in October 1947 Captain Charles Yeager, U.S.A.A.F., became the first man to break the sound barrier while flying the XS-1. (Photo: Bell Aerospace.)

Follow-on aircraft to the XS-1 was the X-2, also from Bell and built of stainless steel. It is seen here gliding back to base after a high-speed run. The X-2 raised the highest speed achieved by man to 2,148 mph in 1956, although it did not set an official record. (Photo: Bell Aerospace.)

First record-breaker after the Fairey Delta 2 was the McDonnell F-101A Voodoo interceptor, 'Fire Wall', and also the first twin-engined record-breaking aircraft since the Meteor. This photograph shows 'Fire Wall' returning to Edwards Air Force Base after a flight over its usual haunts in California. (Photo: McDonnell Douglas.)

Chapter 11

FIGHTER SPEEDS AFTER WORLD WAR II

Towards the sound-barrier – supersonic fighters and bombers – Mach 2.0 – lightweight fighters – Mach 3.0.

A multi-mission combat aircraft of advanced design which could provide the teeth of the Armée de l'Air during the late 1970s. This is another variation on the Mirage themes, but with two engines and near Mach 3.0 capability. An alternative, and more expensive, design incorporates variable-geometry. (Photo: Avions Marcel Dassault.)

A world of difference in the situation of military aviation existed after the two world wars, just as the whole approach to the subject of speed differed. While the fighter was still to represent the peak of military air speed at any one time, the gap between the fighter at its fastest and the absolute speed record was to be dramatically reduced as the result of repeated successful attempts on the record by prototypes of new fighter aircraft. This was in direct contrast to the inter-war years, when only the Messerschmitt Bf.109R was directly related to the fighter of its day, and that was a specially modified aircraft.

Record attempts by modified fighter aircraft and by special high-speed aircraft built for research purposes were not to disappear altogether, but they were not to have a monopoly of the record. Frequently they were to be cut down to size when their records were broken by more or less standard aircraft.

Nor did the lull of the immediate post-World War I period repeat itself in quite the same way, although the demands of a short-lived disarmament programme had their effect, and in the United States production of military aircraft was but a fraction of the peak wartime effort, with only 1,669 military aircraft built in 1946.

A driving force was the chance to exploit the technology of the turbojet, which by the end of 1945 was already raising the absolute speed record out of the reach of the piston engine. Every air force and aircraft manufacturer of any importance was involved with the turbojet, and a succession of new designs appeared throughout the late 1940s after being developed as quickly as possible. In spite of this, the increase in operational air speeds was remarkably steady during the early years of the turbojet, and it was only later that the first leaps came.

A feature of the period was the way in which the turbojet did not reign supreme in every respect. Its shortcomings in range and warload during its early years encouraged the development of advanced piston-engine types, such as the Grumman Bearcat, Northrop Black Widow, de Havilland Hornet, Hawker Sea Fury, Blackburn Firebrand, and Supermarine Spiteful, as well as improved versions of the better World War II types, such as the Lockheed Lightning, North American Mustang, Republic Thunderbolt, Supermarine Spitfire, Hawker Tempest and Fairey Firefly as well as, in Spain, the Messerschmitt Bf.109, fitted with Rolls-Royce Merlin engines!

Such aircraft were not considered to be obsolete or even obsolescent. They could do many jobs better than the turbojet of the day and were often only a little slower. The situation was not so short-lived either. During the Korean War, the radial-engined Hawker Sea Fury proved itself to be a match for the Mikoyan-Gurevich MiG-15 jet fighter. Many also thought at first, although incorrectly, that the turbojet would be unsuitable for aircraft carrier operations, ground attack or night fighter duties, for long-range reconnaissance or for operations from anything but the best airfields with long concrete runways.

Sharing the prevailing attitudes, the U.S.A.F. spent some time after the war in evaluating a piston-engined reconnaissance aircraft of advanced design by Howard Hughes, and its performance in relation to that of turbojets fully justified their doing so.

Although the trend which started during the 1930s – towards extending the range of duties possible for any aircraft to carry out – continued, there was also a trend towards some greater specialization. Of course, the different types might still be based on one basic design of aeroplane. It was no longer enough to have just fighters, for there had to be interceptors as well; there also had to be ground attack aircraft as well as fighter-bombers, and interdictor-bombers as well as differing sizes and ranges of bombers. This was part of the increasing complexity of warfare, but fortunately, the process started to reverse during the 1960s, when fewer types were expected to undertake the maximum variety of duties. One result was that, in spite of the often record-setting turn of speed of many aircraft, range and warload came to be of at least equal importance.

In such circumstances it is not surprising that more interest was paid towards such speed records as that over the 100-kilometre closed circuit, although even so this never had the popular appeal accorded to the absolute speed record. Perhaps this was a reflection of the massive post-war decline in public interest in the whole subject of air speed records.

In the end, the turbojet was to prove itself to have an unrivalled ability to meet the demands placed upon it, allowing aircraft to fly faster, higher, farther and heavier than the piston could ever have made possible; it even displaced the turboprop, which for military applications was reduced to rôles such as maritime-reconnaissance and air transport. This left the Westland Wyvern as the only turboprop strike aircraft to enter service and was the Royal Navy's main carrier-borne strike aircraft of the early 1950s. French and American turboprop strike projects were either cancelled or converted into something else before entering service. This was the case with the Breguet Alize anti-submarine aircraft. Thus, in post-World War II military air speed, there was no turboprop phase to compare with that encountered in civil aviation.

At first, the undoubted leader was the Gloster Meteor, of which the Mark IV version entered Royal Air Force service during the late 1940s. It incorporated some of the modifications applied to the air-speed record versions. In its final form during the early 1950s, the Meteor could manage speeds of up to 640 mph, but naturally the aircraft was not left with any monopoly of service speed.

Britain's second turbojet, the single-engined and twin-boom de Havilland DH.100 Vampire was slower than the Meteor but in certain ways a more useful aeroplane. Perhaps this is best illustrated by the fact that it was the first jet aircraft to operate from an aircraft carrier. By the late 1940s the Vampire had been developed into the 650 mph Venom and Sea Venom.

Even though America's first operational jet fighter, the Lockheed F-80 Shooting Star, took the world air-speed record away from Britain and the Meteor, it was probably a comparable aircraft in terms of speed rather than a faster machine – but in the Korean War it proved to be more manoeuvrable. The piston-engined aircraft already mentioned were contemporaries of these turbojet aircraft and had speeds of well over 400 mph.

Russia's early jet aircraft from the Yakovlev and Mikoyan-Gurevich design bureaux were undoubtedly inferior to their British and American rivals. Still lagging miserably in technology, the Russians had grabbed whatever came to hand while invading Germany towards the end of World War II. They wedded copies of British engines to airframe designs which were basically German and hoped for the best! The results of Ivan's optimism during the early days are not known, for little was ever seen of the Yak-15 and MiG-9, although these were supposed to have entered service. The MiG-15s' Korean War performance was less than brilliant, but noticeably better in the hands of the Russian pilots than in those of the North Koreans.

In attempting to do their best for the then seemingly limitless thrust which could come in due course from the jet engines, the aerodynamicists tried a number of unusual solutions which were mainly linked to specific military needs, such as fighter, reconnaissance or bombing duties. Much of this effort was devoted to finding a suitable supersonic shape.

A number of designers put their faith in the tailless concept, of which the most famous representatives were the de Havilland DH.108 and the Northrop X-4 Bantam, but these and other similar aircraft, including the large Northrop YB-49 heavy bomber, were not really successful. Few tailless aircraft entered operational service, if one excludes the deltas, but the experimental tailless aircraft were not delta-wing designs and had more in common with the World War II Komet. A more promising aircraft was the Bell X-4, with variable-sweep and variable-camber wings, which provided useful background for later designs. The Americans also built the experimental Chance-Vought 173 and CV-XF5U-1 'Flying Flapjack' fighters with large circular wings and twin fins, but these did not live up to their promise of a combination of helicopter and fighter characteristics.

A British aircraft, the Saunders-Roe SR/A1, was the world's first and only jet-powered flying-boat fighter (equipped with two Metrovick axial-flow

L

turbojets), but this project was finally cancelled after successful test flights.

Fighter speeds were at this time firmly in the 600–650 mph range, but this was also true of carrier-borne jet aircraft, which it was now realized were a definite and practical proposition. For the United States Navy the first carrier-borne jet fighter was the McDonnell FH-1 Phantom, and this was soon joined by the related McDonnell F2H-2 Banshee and by Grumman's F9F Panther, which had the distinction of being the first United States jet fighter to see combat during the Korean War. Experimental developments of the early fighter designs intended to test the benefits of the swept-wing included the McDonnell XF-88 and the Lockheed XF-90. The North American XF-86 was modified from its original straight-wing to the swept-wing F-86 Sabre, capable of more than 650 mph and the fastest fighter aircraft of the Korean War. Standard versions of the Sabre took the absolute speed record to the 700 mph mark.

British jet fighters started to lag behind a little at this stage. The carrier-borne Supermarine Attacker was, at best, only an interim aircraft while the fast and extremely manoeuvrable straight-wing Hawker Sea Hawk was developed. A swept-wing development of the Sea Hawk, the trans-sonic Hawker Hunter, encountered development problems and did not enter service until almost the mid-1950s, by which time the Americans were concentrating on successors to their F-86 Sabres and Republic F-84 Thunderjets.

Early jet bomber design did not lag far behind that of the fighter. The English Electric Canberra medium jet bomber was capable of 550–600 mph and the then relatively heavy Boeing B-47 Stratojet, a six-engined aircraft related to the Boeing C-135 and 707 series of transports, had a similar speed range. This made the high flying jet bomber a formidable threat to fighter defences based on aircraft which were only marginally faster. But from such challenges does progress result.

With this in mind the Americans and the British put a considerable amount of effort into developing supersonic aircraft. At first many of the designs were merely trans-sonic, able to exceed the speed of sound in a shallow dive. The Sabre could do this, as could the Hunter and its contemporaries (including the de Havilland DH.110, a twin-engined development of the twin-boom Vampire-Venom series. The all-weather Gloster Javelin – the world's first operational delta-wing warplane – the French Dassault Mystère and the Russian Mikoyan-Gurevich MiG-17 were a little slower.

Although the British Supermarine Swift was intended to break the sound barrier in level flight, the first aircraft to be easily and truly capable of this performance was the North American F-100 Super Sabre, which set the first supersonic air-speed records. During the mid-1950s other aircraft with a similar performance were put into service with the major air forces, such as the

Dassault Super Mystère and the Mikoyan-Gurevich MiG-19 with maximum speeds in the region of 800–900 mph. Strangely, after the Swift's premature withdrawal from R.A.F. service, no British design in the 800–900 mph range appeared. The Americans even managed to get this type of performance on to the aircraft carriers for the early 1960s in the form of the Ling-Temco-Vought F-8 Crusader.

The Super Sabre was soon joined in U.S.A.F. service by the operational development of Convair's XF-92 delta-wing experimental aircraft, the F-102A Delta Dagger, and throughout the late 1950s the delta-wing was very much in vogue. In spite of experimental work with Boulton-Paul and Avro designs, and after putting the first operational delta-wing aircraft into R.A.F. service, the British lost interest in this concept, and the only other delta-wing for the Royal Air Force came in 1959 with the 660-mph Avro Vulcan four-engined long-range bomber – one of three designs in this category to deliver Britain's nuclear deterrent.

Undoubtedly the greatest loss to Britain was the neglect of the potential inherent in the record-breaking Fairey Delta 2 design, which had increased the world air speed record in 1956 by an unprecedented margin, which has not been matched since. The French immediately started work on their Mach 2.0 Mirage III series, doubtless inspired by the success of the British aircraft. Before this, the United States Navy had obtained supersonic capability from production versions of the Douglas F4D Skyray, which was an earlier record-breaking aircraft.

The delta-wing, and such related forms as the so-called double delta, eventually led to a number of Mach 2.0 designs in addition to the Mirage series, and these entered service during the early 1960s. Prominent amongst these were the Swedish SAAB J-35 Draken (or dragon) and the Russian MiG-21. The Americans too built a development of the Delta Dagger, initially with the designation of F-102B, but eventually entering U.S.A.F. service as the Convair F-106 Delta Dart.

However, there was not a monopoly for the sleek delta aircraft as speeds approached Mach 2.0. McDonnell produced the F-101 Voodoo interceptor, in which the lines of the later Phantom II could be vaguely discerned, along with an end to their list of supernatural wierdies! Republic produced the F-105 Thunderchief with near Mach 2.0 performance, and Lockheed pushed speeds over the 1,400-mph mark during the late 1950s with their F-104A Starfighter, although operational aircraft were a little slower. A match for all of these aircraft for the Royal Air Force came with the English Electric (later British Aircraft Corporation) Lightning interceptor, which scored over many of its rivals in having a considerably higher rate of climb and much longer range.

Range was in fact a considerable problem with many of the Mach 2.0 jets, and the French and Russian designs fared particularly badly in this respect. The Russians had to accept a marked reduction in speed with the Sukhoi Su-9, their longer-range interceptor which to this day operates as a back-up to the MiG-21 strength.

Bomber speeds were not allowed to fall too far behind those of the fighters and interceptors in the major air forces. The pattern was set by Convair's delta-wing, four-engined B-58A Hustler, which established a New York to Paris speed record soon after its entry into U.S.A.F. service during the early 1960s and brought Mach 2.0 to the bomber rôle. The United States Navy also managed to obtain high bomber and reconnaissance performance from its North American A-5 (later RA-5) Vigilante, and the French solved their problems of delivering their nuclear weapons by the simple expedient of 'blowing up' (in the sense of enlargement rather than destruction) the Mirage III, and adding a second engine and a navigator/bomb-aimer, thus producing the Mirage IV Mach 2.0 nuclear bomber.

Although designated as a fighter the General Dynamics F-111 variable-geometry aircraft was largely intended for bomber duties at low-level, although plagued with teething problems. A less troublesome aircraft was B.A.C.'s TSR-2, which was a rival for the F-111 in the Mach 2.0 low-level attack rôle, but this aircraft was killed by its political enemies while no fewer than thirty were in various stages of completion. Thus the Royal Air Force was left to spend a decade casting around for a replacement. An Anglo-French project for a variable geometry aircraft came to nothing, while it remains to be seen whether the so-called multi-rôle combat aircraft, the Anglo-German-Italian Panavia 200 Panther, can live up to its planned Mach 2.0 plus strike- and interception-rôle.

In fact this need to get one basic aeroplane to perform as many different duties as possible became an important consideration in military aircraft design throughout the 1960s. Perhaps no aircraft has come nearer to satisfying this demand, and certainly few did so as well as the McDonnell F-4 Phantom II. Originally designed as a United States Navy Mach 2.0 plus carrier-borne interceptor, the record-breaking Phantom II was soon adopted by many air forces and used for fighter-bomber and bomber duties as well. Later versions have been specially adopted for reconnaissance and even tanker purposes. Performance is in fact nearer Mach 2.5 than Mach 2.0. Certainly, Phantoms will be in frontline service for many years yet.

Many bombing needs were being met during the mid- and late-1960s by subsonic low-level aircraft, able to fly under defensive radar nets, and often carrier-borne. Good examples of these aircraft included the British Hawker Siddeley Buccaneer and the American Grumman A-6 Intruder. The argument for such aircraft is that supersonic speed, difficult to obtain at low level and unnecessary, could be dispensed with. Obviously such useful and rugged aircraft could not establish a speed record of any kind.

Rather more controversial was the decision to scrap a British supersonic vertical take-off aircraft during the early 1960s. This was the Hawker Siddeley P.1154. Instead building switched to the subsonic Harrier. Although it may be many years before vertical take-off aircraft can match the speeds of their conventional rivals, Mach 2.0 capability is likely from the second generation of such aircraft during the mid-1970s, including projects under development by North American Rockwell and others.

In all this progress, much of it concentrated into the 1960s, the aircraft became a highly complex and costly piece of equipment, often out of reach of the smaller nations. Pilot training was also a problem, since suitably qualified candidates for training could not be provided for every air force, and the sophistication of many aircraft was also wasteful for certain tasks even in the largest and most advanced air arms. The result was a marked and growing need for simplified combat aircraft, since the needs of the smaller powers could not necessarily be met adequately by second-hand equipment.

This 'lightweight' fighter field was one in which the French scored immediately with the Mirage III, and with its fighter-bomber variant, the Mirage V. The simple air-frame was an immediate advantage and assisted the manufacturer, Dassault, in providing a variety of equipment fits to meet the demands of different customers. This also meant that the simplified versions of Mirage were also the fastest of the aircraft in this field with Mach 2.0 capability.

In America the approach was different. A number of specially designed projects appeared and entered service with the U.S.A.F. and U.S.N. for certain duties on which other aircraft would be wasted. The most significant of these has been the Northrop F-5A Freedom Fighter, and its twin-seat variant, the F-5B, which can manage a speed of not much less than 1,000 mph (Mach 1.5), and has the advantage of twin engine reliability. A development of the F-5B, the T-38 Talon, provides the U.S.A.F. and Luftwaffe with a high performance advanced trainer. It is probably the fastest training aircraft in service, apart from conversion trainer versions of the more sophisticated fighters. The F-5A can be used for fighter or fighter-bomber duties.

A slower fighter-bomber type also adopting a basically simple approach, and definitely subsonic, is the Douglas A-4 Skyhawk. The Skyhawk is really an expression of the continuing need for aircraft of the Sea Hawk and Panther mould for carrier-borne operations, particularly from the smaller U.S.N. carriers and those of the Royal Australian Navy. Some Skyhawks have gone into air force service, however, in Israel and New Zealand.

Other aircraft, developed earlier, such as the Fiat G.91 fighter-bomber, can really be dismissed as belated attempts to introduce a heavier load-carrying capability with the jet. They failed to compare with the slightly later Ling-Temco-Vought A-7 Corsair II. Aircraft of this type are generally subsonic. The Mirage V and the Mach 1.5 Sukhoi Su-7B are perhaps the only exceptions.

Perhaps an extension of this need for simplicity can be found in the attempts by India and Egypt to produce their own national aircraft designs, although to describe either as national is over-generous as the talent of leading German designers was called upon. Not only have such aircraft contributed nothing to the development of military air speed, but their contribution to their respective national air powers has been minor at best. The Egyptian project normally gives its demonstration runs behind a tractor!

A forerunner of what might best be described as the present generation of fighter aircraft was the Lockheed YF-12A and its related reconnaissance aircraft, the SR-71. The YF-12A's speed record of 2,070 mph, and its tremendous range for an aircraft of such a speed, could have made it a significant addition to the U.S.A.F.'s inventory, as indeed the reconnaissance version did become. Neither version was built in substantial numbers, however. The U.S.A.F. wanted something more manoeuvrable for fighter work, and only a handful of reconnaissance aircraft were needed at a time when the use of satellites was becoming widespread for this work.

It was from this background that the concept of the air superiority fighter was born.

The first of these aircraft to enter service has been the Mikoyan-Gurevich MiG-23, with Mach 3.0 capability, although not much more is known about the aircraft other than that it has been introduced to service with the Soviet air arms, and had been operated by Russian pilots over the Suez Canal Zone. Its twin-fin design, illustrating the problems of directional stability encountered as speeds increase so dramatically, is probably the result of a great deal of experience with the Sukhoi E-66, E-166 and E-266 research aircraft. At the same time a bomber designed by the Tupolev bureau, and N.A.T.O. code-named 'Backfire' of similar speed is entering service in the Soviet Union, giving that country a brace of advanced aircraft projects, albeit to the neglect of much else.

In an attempt to maintain her position, the United States has developed two air superiority fighters; the carrier-borne Grumman F-14A Tomcat, with variable-geometry, and the U.S.A.F.'s McDonnell Douglas F-15 Tiger. They have performances comparable to the MiG-23 and perhaps greater reliability. An advanced bomber design to counter that of the Soviet Union is the North American Rockwell B-1.

Such very fast manned aircraft are necessary because of the shortcomings of guided missile systems and the call-back advantage of the manned aeroplane. Those who have persistently believed in the missile solution to defence problems have been repeatedly proved wrong, although it is still questionable whether the latest generation of fighters will ever be superseded by another and even faster generation, with bombers to match.

What is also clear is that the West has allowed the United States to assume full responsibility for producing aircraft capable of matching those from the Soviet Union. The only European country which has tried to maintain an advanced military aircraft production facility has been France which wisely moves forward a step at a time on a low cost basis rather than making expensive leaps forward. An interim design, the swept-wing, and strictly non-delta, Mach 2.5 Mirage F.1, will lead in due course to a Mach 3.0 design. Two designs have in fact been proposed, including a variable-geometry design and a cheaper fixed-wing design, both taking their ancestry from the variable-geometry Mirage G.8 and having two reheated turbojets. There can be little doubt that one of these will enter production, enabling France to follow the United States and the Soviet Union to Mach 3.0, belatedly but nevertheless in the confident state of being able to supply the country's military aircraft needs without the political restrictions and costs of buying from abroad.

Other European aircraft, such as the Viggen from SAAB and the Panavia Panther, are but Mach 2.0 plus at best, while the Jaguar is merely a Mach 1.5 aircraft, built jointly by B.A.C. and Dassault-Breguet for ground-attack duties. German plans for a Dornier-built air superiority fighter of canard configuration attract little attention. During the period after World War II successive German Governments and the country's aircraft industry have become known for unfulfilled flights of fancy.

For those who cannot keep abreast with the Mach 2.5–3.0 developments, either because of lack of inclination or for economic reasons, the United States is developing a new generation of lightweight fighters, with speeds generally around Mach 2.0. An extensive programme of evaluation will decide whether the General Dynamics YF-16 or the Northrop YF-17 is selected. Whichever aircraft is chosen, it will enter U.S.A.F. service as well as that of America's poorer allies.

Such aircraft can only be a side issue in terms of speed. Speed will always be important for warplanes, although it must be tempered with endurance. In terms of speed too, progress is accelerating, if the trend of the post-war period continues. So often this has been coupled with other improvements. These are perhaps the keynotes of the post-war period in aviation: greater progress and a tendency towards all-round progress. The inter-war speed machine was far removed from the fighter, even though the fighter was supposed to represent the peak of practical aviation development, and such speed machines were good for little else than an attempt on the record or, if they managed to stand the strain, a racing event.

Intended to be a Mosquito development for use in the Pacific, the sleek de Havilland Sea Hornet arrived too late for combat duties aboard the Royal Navy's aircraft carriers, and was eclipsed in service by the manufacturer's jet fighters, the Sea Vampire and the Sea Venom. (Photo: *Aeroplane*.)

One of the most widely used American jet aircraft of the post-war period was the Republic F-84 Thunderjet, which served with many of America's allies, particularly before the arrival of the Sabre. The aircraft also saw action in the Korean War. (Photo: Fairchild-Hiller, Republic Division.)

A swept-wing development of the Thunderjet for reconnaissance duties was the RF-84F Thunderflash, also serving with many Western air forces. (Photo: Fairchild-Hiller, Republic Division.)

The first delta-wing aircraft to enter operational service with any air force was the Gloster Javelin, a two-seat, twin-jet, all-weather interceptor for the Royal Air Force. Although a subsonic aircraft, the Javelin soldiered on in the Far East and Zambia until the mid-1960s.

The supersonic Republic F-105 Thunderchief, with Mach 2.0 capability, bore the brunt of the American air involvement in the Vietnam War for many years, proving to be a versatile and reliable machine. This is an F-105D operating in the fighter-bomber rôle. (Photo: Fairchild-Hiller, Republic Division.)

Opposite: Full supersonic capability arrived with the Convair B-58A Hustler jet bomber, which entered U.S.A.F. service during the early 1960s. It is now in reserve. However, Hustlers set a number of point-to-point records during their operational period, including Los Angeles to New York on 5 March 1962 in just over two hours, flying at an average speed of 1,216 mph, and Washington to Paris on 26 May 1963 in three hours, thirty-nine minutes, flying at an average speed of 1,050 mph. (Photo: General Dynamics.)

Above: The Soviet Bloc's Mach 2.0 interceptor is the Mikoyan-Gurevich MiG-21. It is markedly shorter on range than the Lightning or Phantom. Two Polish aircraft are shown here in formation. (Photo: Polish Air Force.)

Opposite, below: The increasing cost of sophistication resulted in a new breed of low-cost, or light-weight, fighters such as the Northrop F-5A Freedom Fighter, shown here taking off with its advanced trainer development, the T-38 Talon. (Photo: Northrop.)

Below: Sweden has consistently been able to provide most of her own fighter needs, with few exceptions over the years. The Swedish interpretation of the Mach 2.0 interceptor is the SAAB-35F Draken, or Dragon, shown here in this three-view general arrangement drawing. (SAAB.)

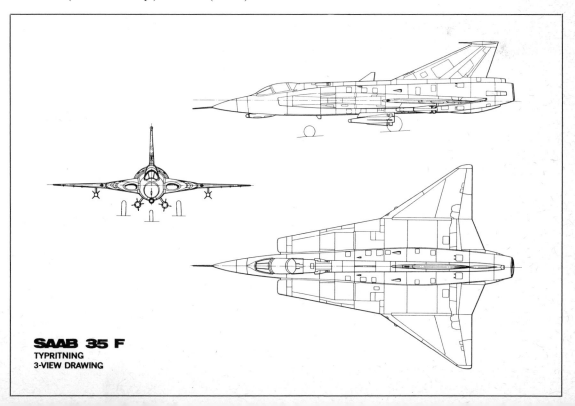

SAAB 35 F
TYPRITNING
3-VIEW DRAWING

A ground-attack development of the Mirage series, the Mirage V is another important warplane, with Mach 2.0 capability while flying 'clean'. This example belongs to the Lebanon. (Photo: Avions Marcel Dassault.)

Left: A British Aircraft Corporation Lightning Mk.6 interceptor in the air near Blackpool. Capable of speeds in excess of Mach 2.0, the Lightning also has a very good rate of climb and a longer range than most interceptors in its class. The range can be extended by fitting wing over-tanks. (Photo: British Aircraft Corporation.)

An artist's impression of the Panavia 200 Panther multi-rôle combat aircraft, which is being developed by the United Kingdom, West Germany and Italy for their air forces. The Panther has variable-geometry wings, and maximum speed should be in excess of Mach 2.0. (British Aircraft Corporation.)

174

Like Boeing, Dassault get full value from every development, driving home the truth in the saying that it is 'better to over-develop than to over-design'. This is a flight of Mach 2.5 Mirage F.1 interceptors. (Photo: Avions Marcel Dassault.)

America's answer to the MiG-23, the McDonnell Douglas F-15 Eagle air-superiority fighter is entering U.S.A.F. service. A Mach 3.0 aircraft, the Eagle has a thrust-to-weight ratio of greater than one-to-one, giving excellent take-off, climb and manoeuvrability. The aircraft is fixed-wing, with all-moving tailplane and twin-fin, an increasingly common characteristic for very fast aircraft resulting from the need to provide adequate directional stability. (Photo: McDonnell Douglas.)

Chapter 12

THE BENEFITS

Aircraft designed from the record holders – the technological fallout.

There was a time when the voyage from the United States to Europe took four or more weeks, depending on wind and tide. Farther afield, the greater chance of becoming becalmed or falling prey to natural and man-made dangers made any prediction about the duration of a voyage still less certain. Travel overland was slow and uncomfortable and subject to a number of uncertainties, not all of which were geographical or meteorological. First the railway and then the steamship gradually changed this picture. The trans-Atlantic voyage, for example, was reduced to four-and-a-half or five days. Improvements in passenger accommodation enabled the railways to satisfy a variety of needs and tastes, and at sea the voyages of the largest liners became glittering social occasions, providing one didn't travel steerage!

Yet in spite of all this, with all the advantages of stabilization and heated swimming pools at sea, and electrification on the railways, by the early 1960s more passengers were travelling by air across the North Atlantic than by sea, and the railways of Europe were dependent on increasingly heavy subsidies. Taking the major capitals of Europe, probably as many aircraft left for the United States and Canada in one day during the early 1970s as there would have been ocean liner voyages in three or four months. Some of these aircraft could carry a quarter of the passenger load of the largest liner. For passenger traffic the railways in North America were forced back into the dormitory areas of the large conurbations, while the air traveller was able to forget a timetable – so frequent were the services.

The social implications of technological change are far more telling and provide a better yardstick for measuring change than any other.

Every innovation brings its social side effects. During the nineteenth century, apart from establishing large new industrial centres for their own needs, the railways assisted the expansion of existing centres by moving raw materials and workers easily over distances which would previously have been a barrier to development. An English clergyman, Thomas Cook, organized the first day excursions by rail, starting a trend which in due course saw English fishing villages grow into fashionable Victorian holiday resorts.

The growth of air travel and its ever-increasing speed and frequency not only oiled the wheels of trade and diplomacy, substituting the personal visit for the written or telegraphed communication or local representative, but also took pleasure-seekers even farther afield. The fishing villages of the Iberian Peninsula and of the rest of the Mediterranean area, and their equivalents in the New World, became the Blackpool, Brighton and Atlantic City of the late twentieth century. Local populations in hitherto poor and remote areas suddenly found an affluence and bustle thrust upon them which must be considered a mixed blessing.

M

Obviously, an appraisal of aviation speed is not a sociological survey, but it must seem that the hallmark of the jet age has been not so much the rapid exchange of ideas as the standardization of ideas and the loss of individuality, with one city beginning to look much like any other, regardless not just of country, but of continent too.

The real instrument of change during the late twentieth century had a turbojet or turbofan engine, a swept-wing all-metal structure and a monocoque fuselage, while the undercarriage was retractable, the controls power-assisted, and to aid take-off and landing there were such devices as flaps and slats. A swept-wing is not essential for flight at 600 mph, as the Meteor, Shooting Star and Skystreak showed, but, whilst belonging to a much higher speed range, such a design does make travel at such speeds more economic and more comfortable. In every way advanced technology often serves its purpose by assisting at a much lower, or in this case slower, level than that for which it was originally intended. All the features which make a modern airliner different from the wooden biplane with radial engine and fixed undercarriage which served during the early 1920s, owe their existence to the quest for more from the aeroplane – greater range, higher payload and better accommodation, and above all, more speed.

It is not merely a case of moving up to 534 passengers at 600 mph and 35,000 feet over ranges of up to 6,000 miles without refuelling, but also one of moving up to 100 tons of freight. Air freight is not a luxury for those in the remotest areas, on the contrary it is often a necessity, a mundane and normal means of conducting business. Britain's second most important port in terms of the value of traffic handled is London's main airport – Heathrow.

The rapid restocking made possible by air freight means that less capital is tied up in goods in transit or in store, and to this can be added lower insurance and packaging costs making air freight cheaper than slower forms of transport.

The advent of the large jet airliner was thought to be the doom of the airlines, and at first jet passengers paid a surcharge. But in the end, high utilization, good productivity and safety, and a rising standard of living meant that fares fell in relation to the cost of living. Airlines were able to sell their old aircraft fairly easily to other airlines which put the remainder of their life to economic use on even lower cost charter flights, stimulating the economy of many parts of the world and taking the people of the developed countries to those places where their money would stretch further. The process continues.

Statistically, the British Overseas Airways Corporation carried just 10,000 passengers a week during 1945, before the formation of British European Airways, while the successor to both airlines, British Airways, carried some 15,000,000 in 1972. The growth implications cannot be ignored.

The first record-holding aircraft, designed and built by the Wright brothers, set a pattern for many other designers to copy, modify, and then, using the results of this work, to produce their own original designs. There were other designers who followed their own ideas right through from the start without obviously copying the brothers. But, even though some of these aircraft actually managed to fly, for most the sure way to the sky was via the well-trodden path of Wilbur and Orville Wright. Thus it can fairly be said that even such early aircraft provided technological and military fall-out, although the social implications may have been less marked.

In its structure, the Monocoque Deperdussin set an example for other aircraft to follow, although the monoplane concept itself was to fall into disrepute, and for many years offer little advantage over the biplane. It was in fact in their streamlining that the French and American biplane-record holders of the 1920s contributed to aviation, although they also showed the merits of higher wing loadings and lighter structures.

The unsuccessful Supermarine S.4 led to the successful S.5, S.6, and S.6B, and these aircraft were the direct ancestors of the Supermarine Spitfire fighter of World War II, with their low-drag monoplane structures and reliable and high performance water-cooled in-line engines, happily also possessing a low drag characteristic. The lessons were not lost on the Germans, and the Italians also used their later inter-war designs as a basis for wartime fighter development.

In many ways, the Messerschmitt Me.163 Komet interceptor of World War II was more the predecessor of the space shuttle, which will also be ramp-launched, than of the American research aircraft of the post-war period; although these also have doubtless contributed towards shuttle technology Not only did the Komet introduce a new concept to air defence, that of the fast climbing interceptor to counter high altitude bombers, but its relationship to the shuttle is much the same as that between the V-2 rocket and the I.C.B.M. of today.

Unfortunately, nothing of commercial significance came from the Komet, the DH.108 or other tailless designs, except the negative finding that perhaps such designs had no commercial value, and that something more suitable should be tried for the de Havilland Comet 1. It was in any case decided that the Comet should not be a small mailplane, but an airliner, although the swept-wing does owe something to DH.108 technology. It was an innovation for air transport, and came at a time when the incorporation of a swept-wing into an aircraft design was by no means a foregone conclusion.

If the turbojet itself was a result of the need to go faster, then many of the engines of today must still owe much not only to this early impulse of Whittle and Von Ohain, but to many of the record-breaking jet aircraft. The Rolls-Royce Avon, which, with reheat, powers the B.A.C. Lightning and the SAAB-35 Draken and, without reheat, powers the Hawker Hunter, was used and underwent development as the power unit for the record-breaking Hunter, Supermarine Swift and Fairey Delta 2. The latter aircraft was subsequently modified to test Concorde aerodynamics.

Not all research aircraft were built for speed. Some, such as one Handley Page design, were designed to test low speed handling characteristics. However, such aircraft were built as a result of the need to utilize the results of high speed flying and to see how the aerodynamic concepts tried and proven for high speeds would behave at the low speeds necessary while an airliner is on the landing approach to an airport.

A great deal of work on materials resulted from the speed record aircraft. Metal replaced wood, and special alloys were developed to cope with the heat generated by high speed flight. Again, the negative result was sometimes the benefit. Bristol, in building the stainless steel 188 research aircraft, discovered that this was a difficult and therefore costly material to work. The result was the decision to build the Concorde supersonic airliner with conventional technology and materials. Even so, this has meant an estimated cost so far of well over £1,000 million, divided equally between the United Kingdom and France, and including post-introductory developments and testing, but excluding another £20 million or so spent by the British Government on Concorde-related research with the Handley Page and Fairey deltas.

The American attempt to build a faster supersonic airliner involved introducing new technology, and the problems and costs involved eventually forced the United States Government and its contractors, Boeing, to abandon the project. Since costs considered too steep by the Americans would be out of the question for the combined resources of the United Kingdom and France, the decision to build Concorde using existing technology and materials, and accepting in return a speed restriction of a little more than Mach 2.0 instead of Mach 2.5 plus, has been vindicated. The Bristol 188 therefore justified its existence, even without making the once hoped-for attempt on the official absolute speed record.

Speed therefore has not always been an end in itself. It is, of course, born of man's impatience and curiosity, but in the end the outcome has been of commercial and military importance, particularly relevant during World War II with the Spitfire, and during the post-war period with the attempts on the record by versions of standard fighter designs. The need to maintain adequate defence is a good reason, if not the best reason, for pushing the speed of fighter aircraft forward all the time; and since attack is the best and ultimately a major part of defence, progress with the design and speed of bombers has also been necessary. However, for the man in the street, for whom freedom is perhaps a difficult concept to grasp, unless he is deprived of it, the real significance and objective of greater speed in the air must be the closing of the world's distances.

Increasing civil aircraft speeds were accompanied by a growth in aircraft size and range and in a marked improvement in passenger comfort. Changing aircraft shapes can be judged as fairly from the interior of an airliner as from its exterior. Four illustrations of the standards of comfort offered by a major airline and its predecessors amply illustrate the development of civil aviation since the mid-1920s. First, an Imperial Airways Argosy trimotor (with single seats on either side of a narrow aisle, box-like structure and basketwork standards of comfort, which could maintain a 90–100 mph cruising speed after its introduction to service on European routes in 1926. Secondly, bridging the Imperial-British Overseas Airways Corporation gap, is a de Havilland Albatross of 1938, with a 'railway-ish' interior, albeit with a superior standard of comfort and still fairly narrow fuselage (although seating on either side of the aisle has doubled and the structure is rounded). This was a pleasant way of cruising at more than 200 mph! A luxurious view of first-class seating on a 400 mph Bristol Britannia 312 of the late 1950s is the third picture, contrasting with the triple chairs on either side of the gangway of the aircraft in charter service after retirement from B.O.A.C.'s North Atlantic routes. Finally, 'eat it all darling or you don't get off!'-air travel for the masses in a British Airways Boeing 747 'jumbo', although it could almost as easily be an airbus, with double, quadruple and then treble chairs abreast in economy class, separated by two fairly wide aisles. 600 mph is the cruising speed for the seventies. Ten-abreast seating would usually be the order of the day in a charter airline 747 or airbus. (Photos: British Airways.)

Chapter 13

THE FUTURE

Man is never content unless travelling towards the next horizon. Of this, speed is just one aspect; exploration and scientific research are others.

The speed record started as an event of sensational and general interest, with all the novelty of a long awaited new invention still fresh. It was rapidly caught up in the gay and fashionable Rheims Aviation weeks, an aeronautical parallel with Royal Ascot if ever there was one. From this glorious peak started a decline passing after World War I, through a sporting phase of air races, and then becoming more or less routine, and sometimes dull, scientific and military research after World War II.

In many respects, such progress really only means that the record followed the fashions and attitudes of the times. The gay Edwardian era, with civilization and fashion still centred around France during the period before World War I, rising to what French historians term 'La glorieuse année' in 1913, was a like period for aviation, 1913 being a vintage year for the speed record. Between the wars, the Schneider Trophy and other races, such as that for the Pulitzer Trophy, fitted into a scene which also included Le Mans at its heyday and Brooklands enjoying its largely inter-war existence. After World War I, whether public interest had become preoccupied with more mundane and material matters or blunted by years of wartime hardship, the approach became very matter of fact and scientific in this field as in everything else. After World War II much of the colour had disappeared, and interest, even when aroused, could not be maintained for long. Before the end of the Apollo lunar landing programme, even that staggering scientific and human achievement was taken for granted!

No one can really be pleased with this tendency to dullness, for dullness it is. The attitude and approach of the individuals concerned reflects this. Lord Brabazon of Tara, who headed the British committee dealing with post-war transport aircraft requirements which was named after him, was one of the first, as Moore-Brabazon, to obtain a British pilot's certificate. He was also a founder member of the Royal Aero Club and on one occasion took a pig for a flight in a balloon, just to prove that pigs could fly! The Hon. Charles Rolls, of Rolls-Royce, lived a very frugal existence, frequently arriving for lunch at the Royal Aero Club and taking only the water, rolls and butter provided free, eventually causing the R.Ae.C. to impose a table charge.

Flamboyant colour schemes are no substitute for personal character. Alberto Santos-Dumont gained fame for flying round the Eiffel Tower in his dirigible No. 6, and for making the first tentative aeroplane flight in Europe, but he was not above landing in the middle of Paris for coffee, and the onlookers would soon rally round to assist in the mooring of his airship. Even in air transport there were individualists before every aspect became regimented, or the colour became artificial and forced. The story is told of a pilot on B.E.A.'s Scottish Highlands and Islands services after World War II who, lacking a uniform because of post-war shortages and airline reorganization, would sit with the passengers in his Rapide until take-off time, and then angrily demand that the pilot should get on with it, otherwise he would fly the plane himself. When, of course, no pilot appeared, he would, in mock anger, stalk into the small control cabin, start engines and take-off to the horror of those who were not regular travellers on the route.

Increasing automation on the flight deck may have contributed to the present lack of individuality. Even those flying 'basic' aircraft on back country routes can feel that in the eyes of some they are perhaps a little less than first rate, and they are sometimes merely concerned with accumulating experience so that they can pass on to something more spectacular. Perish the thought that they might be developing a skill of their own!

Service flying still offers some scope for enthusiasm, although budget-conscious authority tends to take a 'don't bend it and hang your ability in a dog-fight' attitude. However, there must be something left, for example, in the captain of the British aircraft carrier who, in the midst of a Socialist government's assault on the Fleet Air Arm, hung a giant-size 'Fly Navy' banner over the stern while entering port.

Impersonality is the disease of the late twentieth century, but what of the future for the speed record?

Almost certainly, one or the other of the new United States air superiority fighters will try to raise the YF-12A's record, but only the absolute record since neither aircraft could cope with all the speed-with-range records of the Lockheed aircraft.

For the highest speeds, the arrival of an aircraft faster than the X-15A-2 will depend on the development of the space shuttle (designed to carry men and material to orbiting space stations) during the late 1970s, or the 4,000 mph hypersonic transport of the 1980s, if it is ever built. Certainly the shuttle will come and perhaps raise speeds within the atmosphere to double those of the present as it accelerates into earth orbit and the 18,000 mph or so required to catch an orbiting space station. The hypersonic transport is less certain in an age increasingly concerned with the environment and with a shortage of fuel suitable for aviation use. Designs have been produced, however, notably by the famous English designer, Sir Barnes Wallis, who also did early work on the variable-geometry wing and designed the Wellington bomber of World War II.

That something faster than the aircraft built so far will come should be in no doubt. That it will contain even a tenth of the glamour and excitement of its illustrious predecessors is open to question.

The future? A B.A.C.-North American Rockwell design for a space shuttle, recovering applications satellites for maintenance and putting new satellites into orbit. Craft such as these may well become the fastest manned vehicles and would spend most of their flying time within the atmosphere. (Photo: British Aircraft Corporation.)

The future? A model of a possible hypersonic transport, capable of Mach 5.0, with its designer, Sir Barnes Wallis. Sir Barnes has always ranked amongst the most prominent British designers, although it is only in comparatively recent years that he has become so interested in speed. He can take much of the credit for early design in variable-geometry technology. (Photo: British Aircraft Corporation.)

Index

Numbers in italics refer to illustrations